PANICOLOGY

PANICOLOGY

Two Statisticians Explain What's Worth
Worring About (and What's **Not**)
in the 21st Century

SIMON BRISCOE

&

HUGH ALDERSEY-WILLIAMS

SKYHORSE PUBLISHING

Skyhorse Publishing books may be purchased in bulk at special discounts for sales promotion, corporate gifts, fund-raising, or educational purposes. Special editions can also be created to specifications. For details, contact the Special Sales Department, Skyhorse Publishing, 555 Eighth Avenue, Suite 903, New York, NY 10018 or info@skyhorsepublishing.com.

www.skyhorsepublishing.com

10 9 8 7 6 5 4 3 2 1

Library of Congress Cataloging-in-Publication Data

Briscoe, Simon.
Panicology : two statisticians explain what's worth worrying about (and what's not) in the 21st century / Simon Briscoe & Hugh Aldersey-Williams.
    p. cm.
Includes bibliographical references.
ISBN 978-1-60239-644-9 (alk. paper)
1. Psychology--Statistical methods. 2. Skepticism. 3. Fraud. I. Aldersey-Williams, Hugh. II. Title.
BF39.B76 2008
301.072'7--dc22

                              2008046640

Printed in the United States of America

To the next generation—
Marjolaine, Sam, and Roch—
who will live with the consequences of our actions.

Contents

Acknowledgments

Many people confided their favorite scare stories to us as we set about working on this book, and we are grateful to them for their efforts in persuading us to share their fears. Family members dutifully scanned the newspapers for appropriate alarums, gave helpful pointers or at least allowed us space to write. Our erstwhile chemistry teacher, Mike Morelle, surprised us with a fat file of cuttings that he had apparently been keeping on the off-chance that two of his ex-pupils might write this book. Thank you all, and particularly to our wives—Moira and Laure—for their support. A number of libraries and media databases greatly facilitated our searches. We would like to thank especially the librarians of the Royal Society of Arts, the University of East Anglia, and Cambridge University. Huge thanks to staff at the *Financial Times* for supplying inspiration—and allowing Simon time off to work on the book.

We sought out experts who might inform our potential fears. The following were exceptionally generous with their time, expertise, and contacts: John Adams, Tony Allan, David Berube, Petra Boynton, Tracey Brown, Derek Burke, Ian Burnett, Clark Chapman, Mike Clark, Piers Corbyn, Julian Dowdeswell, Greg Durocher, Marie Edmonds, Kerry Emanuel, Andrew Evans, Pamela Ewan, Brian T. Foley, Kenneth R. Foster, Fiona Fox, Elspeth Garman, David Gee, Mike Hulme, Peter Lachmann, Chris Landsea, John Lawton, Stephen Leatherman, Georgina Mace, Jonathan Matthews, Bill McGuire, Jeffrey McNeely, Clare Mills, Julia Moore, Lynne Moxon, Kenneth L. Mossman, Brigitte Nerlich, Lembit Opik, Julian Orford, Stephen Pacala, Hugh Pennington, Raj Persaud, Roger Pielke, Richard Robertson, Jonathan Shanklin, Tom Stewart, Karen Talbot, Chris Tyler, Craig

Wallace, Brian Wynne. We are grateful to all of them. Tony Lacey at Penguin deserves special thanks for his guidance. Alaina Sudeith, our New York-based editor at Skyhorse, worked tirelessly translating our words and figures into fluent American to prove that we are not after all divided by a common language.

Hugh Aldersey-Williams and **Simon Briscoe**
March 2009

Introduction

"This book made me feel sick. It also, equally effectively, made me feel ashamed, despondent and anxious, increasingly disenchanted with our politicians and, above all, guilty . . . this invaluable book."

This review appeared in a serious broadsheet newspaper. The book happens to be about over-fishing, but it could just as well be about global warming, family breakdown, or hospital superbugs. Observe not merely the dire mental turmoil that the book induces, but the reviewer's pathetic gratitude at being brought into this state. As a personal response, it is a touch overwrought, but it is typical of the way we are all increasingly invited to feel about the endless catalog of disasters that are supposed to await us.

Our book won't make you feel these things. With luck, it may even make you feel a little happier about the condition of the world.

Consider bird flu. Throughout the winter of 2005–2006, this was billed as an impending human pandemic that would wipe out a large proportion of the Earth's population. Yet the H5N1 type of the flu virus has led to the deaths of fewer than 300 people worldwide, mostly in Asia where victims had come in direct contact with infected birds.

Still the media is intent on sustaining scare stories like this everywhere it can—and we lap them up.

Some days, of course, not much happens, and even the media is stumped for a scare story. Actually, not much happened throughout the whole of 2006—the year we began work on this book—in 2007, and into 2008. No suicidal zealots flew passenger jets into high-rise buildings. No hurricane destroyed a major city. No killer wave arose casually to sweep away a couple of hundred thousand shore-dwelling souls. No new plague struck. We live in a complex world and we don't want to die. And in general we are winning the battle—we are living longer and more healthily than ever. Every year, death comes a year closer for all of us, meanwhile life gets a little better for many people. So why are we happy to panic about the silliest things?

This tendency is certainly not new. Charles Mackay's *Extraordinary Popular Delusions and the Madness of Crowds*, first published in 1841, cataloged public obsessions with witchcraft, mesmerism, and tulips, as well as fears of annihilation by everything from flooding to chemical poisons. Today, read pedophiles, radiation, and blueberries . . . and flooding and chemical poisons. These topics are not "hard" news or exactly fact, even if they do have a basis in fact. They live as stories because we all love to gossip, hear a tale, embellish it, and retell it. Journalism is industrialized gossip, as Andrew Marr puts it.[1] Once one newspaper's story about something extraordinary, say a killer bee, has gone down well, others follow, rooting out killer bee–related items that would otherwise have gone unreported, or building up killer bee near-misses into full-blown dramas in their own right. The fact is that we love to be scared—which is why many of the topics we examine (and the bees) have their own disaster movies.

These news stories frequently give the impression that life as we know it is about to end. The nature of the threat may change—a wave of immigration, AIDS, rising sea levels, or an asteroid—but the threat is always there. Life is getting harder in the media as costs are cut. Journalists sometimes over-egg their stories to get them on the front page. And sometimes they are just overstretched by the demands of 24-hour rolling news from around the world and being expected to comment in real time on topics they know little about. Mackay notes how an "epidemic terror of the end of the world has several times spread over the nations."[2] It has happened at regular intervals since before the rise of modern science and has continued following the Industrial Revolution. And it seems it is happening still—in some cases changing our behavior and leading to emotional and financial costs.

What's new, then, is not the public's appetite for a good panic story, or the media's willingness to serve one up. It's the role of other agencies. A generation ago, there were concerns in the world and these were reflected in the media, but the tone was different. Individual accounts of gruesome events may have been both more factual and more bloodthirsty than would be allowed today, but the tone would have been sober, and behind it all would have been the reassuring presence of a paternalistic government.

Today, that comfort blanket has been whipped away. Now politicians and government officials seem more likely to add to our sense of panic about a given issue. And even if our public servants don't, then there are plenty of other highly vocal interest groups—scientists, health and safety nuts, corporations and their advertising agencies, non-governmental organizations and lobby groups—who will. Professor Stephen Hawking said in 2007 that "life on Earth is at ever-increasing risk of being wiped out by a disaster such as sudden global warming, nuclear war, a genetically engineered virus or other dangers." Did he believe this or was he just promoting his zero-gravity flight to increase interest in space travel?

Governments have a duty to warn us of dangers that they perceive, but the way they do it today often only seems to add to the alarm. They are keener than ever to regulate for a safer and fairer world wherever there might be the slightest risk to health. An initial, well-intentioned impact assessment can rapidly snowball into a firmly held policy view supported by the full weight of government. Modern-day communications in our globalized world bring us stories more rapidly than ever before. Recent years have seen government campaigns in many countries about terrorism and bird flu which have raised our fears. But we never really learn the genuine extent of the risks; nor are we told what we as individuals can do to reduce them—we are merely told to be alert—that is to say, on the edge of panic.

The courts have also helped to alter our perception of what is a reasonable risk, sometimes in laughable ways. In the more litigious parts of the world (including the United States), the old rule of "buyer beware" has been sidelined—now the seller gets sued. Following a lawsuit, McDonald's has had to print a warning on its coffee cups pointing out that the contents may be "extremely hot." As we witness the death of common sense, children's playgrounds lose their more exhilarating rides and doctors cover their professional reputations by putting patients through unnecessary tests. The end result is a distortion in our fears that overlooks evidence in favor of sensation: We now fear fires more than drowning, even though more people die from drowning, simply because fires make better television.

The classic social scientist's equation has it that the risk of an event is the likelihood of its happening multiplied by the impact if it does. So the risk of being killed by a volcanic eruption or a terrorist attack depends on the odds of the event, the event's magnitude, your proximity to it, your protection against it, and so on. But a more recent formula begins to take account of the way the media and other agencies are raising the stakes, suggesting that risk = hazard × outrage.[3] This is simplistic, but it clearly admits an important new factor. Governments advise against visits to places they judge to be at risk from terrorist attack but are less concerned about volcanic hotspots because only deaths from the former provoke outrage.

The topics we have chosen show how this wide societal network now manipulates our perception of risks. We have selected some global and some local concerns, some that are easy to understand and others for which the state of knowledge is low. All of them have been prominent stories in the media. We have scored each topic using a points system to show how vividly each threat is portrayed in the media, how real we feel the threat is, and how much we as individuals can do about it. Some of these subjects you may be worried about already. To others you may never have given a thought. Our personal and unscientific assessments might prompt you to reassess your own fears. However, by parading so many popular fears between these covers, we hope at least to show that you cannot worry about everything, and also that it is foolish to worry excessively about any one issue.

We have neglected many risks—including mad cows, child abduction, and the Y2K computer scare. We have also left out the things that are likely to get you, such as cancer, heart disease, dementia, or simply falling. We might have discussed the nuclear threat—it has not gone away. We might have dealt with environmental pollution, still a major concern though no longer the public mania that it was a generation ago. Right or wrong, these risks do not give rise to much panic these days, and so we have put them aside.

Examining panics en masse, we begin to pick out common threads not seen when they are considered, as they generally are, in isolation. There is a general difficulty in accepting that natural events occasionally still have the power to overwhelm us. At the same time, there is an almost biblical inclination to blame ourselves for things that may not be our fault, such as new viruses or freak weather events. There is a fear of forces that (we believe) we ourselves have unleashed through our arrogant scientific optimism. There is disbelief at the limitations of medical science, expressed in outrage at the deaths of infants or the presence of bacteria in hospitals. There is a growing distrust of the government hand.

Above all, there is a paradox. Modern life has greatly reduced many of the risks that humankind has to face, and yet it is modern life that seems to spawn most of our fears—fears of chemical, biological, and nuclear war, pollution, terrorism, climate change, and, less directly, fears associated with immigration, aging, loss of cultural diversity, and much else besides.

As we said, the last couple of years were quiet. There were merely the millions of expected deaths from malaria, HIV, poor water quality, war, and car accidents. Searching the Internet, it is almost impossible to discover how many die during a year from the flu—the sort disingenuously dubbed "seasonal," as if there was not a damned thing anybody can do about it—because the figure (as many as 500,000 people) is all but lost amid completely hypothetical death tolls for the bird flu pandemic that did not happen.

We notice spectacular or novel disasters, but neglect familiar killers. This is human nature. But another reason for this is the genuine gap in our knowledge of risks. The media, just like the public, attempts to navigate the daily news flow relating to global warming or the state of immigration, but, again like the public, has no means of knowing what is right or wrong. The length of the lines forming outside branches of the Northern Rock bank in Britain in the late summer of 2007 reeked of poorly informed panic about money, and it is a wonder that the economy took so long to slow down given the doom-laden credit crunch coverage in the following year. And the problem gets worse as time moves on. New and increasingly complex technologies beyond the comprehension of most bring new risks as business may be threatened by Internet-based markets or our health jeopardized by nanoparticles. Breast cancer is bad, but there are serious risks associated with just screening for the disease. In cases like these, what we would like is a quantitative statement of Robert K. Merton's famous law of unanticipated consequences: How great are these consequences compared to the negative impact of the original problem and the positive impact of its technological solution? But of course this figure is seldom calculable. Sometimes, side effects are negligible; other times, they seem greater than the original problem.

Still, we can take heart from what we do know. Not many of the dangers we confront are absolute—very few are likely to kill many people. Most are relative risks—things like eating too much salt that might knock a few years off your life or a flood that might result in the loss of treasured possessions and a tedious insurance claim. And the value of investments can fall or rise. Yet academic studies of happiness suggest that it is the relative risks that matter to us. As the world becomes more complex and we grow better informed about events, we worry more about these relative risks. But how relative are they?

Numbers are the "fact" generator in today's society and the currency in any debate about risk. But they are not all of equal quality—some are manipulated by governments while others are produced by people with a vested interest. Often, proper figures don't exist—they are opinion surveys or come from administrative systems that do not give us data on the definition we want, leading to poor policy and weaker assessment. Yet those who wish to make a point on television or in the newspapers do it using numbers. Sound-bite statistics, sometimes invented and often inaccurate, seize the imagination even if they crumble under close inspection. What does a one-in-a-thousand chance of catching SARS actually mean? Where in the world are you? What precautions are you taking? Are you of a vulnerable age?

Figures are one of the main ways to spread fear. We might like to think that the figures are the hard facts, the irrefutable hard-cooked foundation for the argument, but sadly they are often not. They can be misleading or deliberately distorting. As John Allen Paulos puts it in his 1988 book *Innumeracy*, "Mathematics is the quintessential way to make impressive-sounding claims which are devoid of factual content."[4]

One is reminded of the famous remark attributed (probably erroneously) to Benjamin Disraeli, the British prime minister: "There are lies, damned lies—and statistics."[5] Yet statistical data are better than nothing—they are evidence of something, the starting point for a discussion, a way of understanding society. The numbers are not everything but they can inform analysis and provide the creative impetus needed to solve problems. The only alternative is to retreat into anecdote and hopelessly selective assumptions.

However, the cult of innumeracy remains strong among the public—and the media. One leading newspaper recently announced that there was "about" a 50 percent chance that Europe would have above-average temperatures in some coming period.[6] Indeed. In one of his acts, the late George Carlin invited his audience to consider how stupid the average American is. Then he paused before observing that half of them are even more stupid than that![7]

Although statistics about the past can be dangerous, forecasts about the future are even more dodgy. Questionable data are put in a black box computer model, cranked and spewed out often, it seems, with the sole purpose of scaring us. The "results" have authority because it is experts and academics who do the cranking. Yet history is littered with examples where economists, scientists, and other specialists have gotten their projections fantastically wrong. Furthermore, the same raw data can be made to yield very different projections according to the prejudices of the person cranking the machine and small adjustments made to the model's assumptions.

The difficulty we have in dealing with the numbers that express risk may be a symptom of a wider inability to evaluate risk at a human level. In part, this reflects a deliberate avoidance of unpalatable truths—smokers still may not give up smoking even though they know it will be the principal cause of death for half of them. But mostly, it's down to ignorance. Perhaps you are sitting at home reading this. If so, you probably have no idea of the hazards that confront you right now. Are you more at risk from an airborne infection, a rat chewing through the wiring and starting a fire, or an asteroid crashing through the roof? You have no idea.

We have plenty of evolutionary equipment to help us evaluate immediate danger. Our senses tell us where to tread and what's safe to eat. But even the simple act of crossing a road is not so black-and-white. How can we choose between driving, flying, or taking the train if we want the safest journey? What about radiation, which we willingly accept in the guise of an X-ray but fear otherwise?

Given this uncertainty, it is no surprise to find that people in different countries fear different things. The Swedes worry about dangerous chemicals, the Danes about nuclear power, and the Italians about radiation from their beloved cell phones, even though the risks from each are probably broadly equal in these countries.[8] Worries also change over time. A disturbing recent survey of Australian children found their main fears were being hit by a car, being near a bomb, and being unable to breathe.[9] In the same survey twenty years ago, a trip to the headmaster, catching germs, and falling over came top of the list.

That's a huge shift. Of course, risks change over time. Worries about terrorism and, well, anything from the following pages have replaced our parents' Cold War worries of nuclear annihilation and communism. But, it seems, our perceptions of risk are much more changeable.

It is almost as if we have to be afraid of something, as if we carry about in our heads a bucket of worry that we are compelled to fill with whatever's available. Clearly, different individuals have different-sized buckets. But the important questions are: is the size of *your* bucket fixed, or does it expand and contract according to external circumstances? Does it expand only when there genuinely is more to worry about, or is it swelled by the media, governments, and other interests? Acting collectively or individually, can we shrink our buckets? If so, how? And should we do this?

According to the anthropologist Mary Douglas, writing with the political scientist Aaron Wildavsky in *Risk and Culture,* "people select their awareness of certain dangers to conform with a specific way of life."[10] A society united in fear is more cohesive. One where people fear different things is liable to fragment. Our dinner-table gossip and actions based on fears—such as the rise or fall of house prices and immigration—serve to strengthen or weaken the social fabric. So shrinking our bucket of worry too much may have consequences for social organization.

We live at a time of unprecedented prosperity, mobility, and connectivity. Most of us live at peace in democracies. It is not merely coincidental that we are also witnessing a loss of respect for authority, a fragmentation of society, and rising levels of worry about ever smaller risks. They are a logical consequence of these developments. What we are seeing is people's first uncertain response to greater technological and political freedoms. We are free to choose what we eat from a bewildering range of foods, for example. Governments, NGOs, and corporations pull us this way and that—cheap, organic, low-salt, high-fiber, local or exotic, and so on. But the choice is ours. Sometimes, though, we wish somebody would make it for us.

What can we do? For many panic topics, we can bend the odds in our favor by being aware of controllable threats. We can free our minds by deciding not to worry about others. We can begin to weigh risks and measures to deal with them. For example, some safety measures, such as making cigarette lighters childproof or putting reflectors on trucks, have proved very effective in terms of saving lives, while others, such as restrictions on hazardous waste, have merely proved expensive.

It is quite legitimate to ask whether more overseas aid money should be put into flood defenses in poor countries. If we want to cut road accidents, perhaps we should ban pedestrians from crossing the road while using their cell phones just as we ban drivers. Perhaps we should have a war on obesity rather than a war on terror. If we as voters were a bit wiser, and were not so easily freaked by dreadful or unexpected deaths by means such as hijacking or BSE, we might get better government.

"Men," wrote Charles Mackay, "go mad in herds, while they only recover their senses slowly, and one by one."[11] After reading *Panicology*, you will, we hope, worry less about many of our subjects; in one or two cases, you might be prompted to worry a little more. But you will have begun at least to prioritize your worries. You are on the road to recovery.

Key to scores (from 1 to 5)

Panic:

Risk:

Personal empowerment:

www.panicology.com

1. Sex, Marriage, and Children

Worrying about whether there are too many people on our planet or too few, and how we should relate to them, is at the heart of many panic stories. Former British Prime Minister Margaret Thatcher told us two decades ago that there is no such thing as society, but the families we make—and break—and the increasingly diverse relationships we form, and see others form, cause plenty of concern. And don't forget how we all got here—sex—is it a problem or a pleasure?

The Birth Dearth

"A second baby? Russia's mothers aren't persuaded"
—*The New York Times*

Italian men not helping much around the house is apparently one of the principal reasons why Italian women are producing so few babies.[1] Other reasons for the birth rate of Italian women falling to be among the lowest in the world include the lack of flexible work, a shortage of nurseries, and the poor provision of children's services, in a country where couples have traditionally relied on families for support. A low birth rate might seem an unlikely problem for a predominantly Catholic, child-loving country, but a serious shortage of babies and the prospect of a shrinking population is affecting many developed countries, to such an extent that it could soon threaten their livelihood and viability. Mass immigration, not always seen as desirable in the West, might become a necessity.

The prospect of too few people being a problem is a far cry from the impending "population crisis" that most of today's adults were brought up with. "Too many people in the world?" was the provocative question on the cover of the September 1963 issue of *U.S. News & World Report*—and typical of the genre. Declining death rates, in other words increasing life expectancy, around the world contributed to the more than doubling of the world's population since 1950 to its current 6.5 billion. It is now increasing by a little over 6 million a month, roughly 200,000 every day. The consequences of this growth are enormous shortages of water and fuel, the depletion of natural resources, high unemployment rates, pressure on public services including education and healthcare, increased ill-health, damage to ecosystems, and pollution.

It seems, then, that there are too few people in some areas and too many in others.

Population fluctuations and associated scare stories are nothing new. The highly influential *Essay on the Principle of Population*, written by Englishman Thomas Robert Malthus in 1798, predicted that population would outrun food supply before the end of the nineteenth century. His basic view was that population, if unchecked, increases exponentially, at a geometric rate, whereas the food supply grows in a linear fashion, at an arithmetic rate. Malthus saw the solution to rapid population growth as being "moral restraint"— including late marriage, which paradoxically is one of the key features of the problem now facing many Western countries with low birth rates and declining population. If we failed to embrace such restraint, excessive population growth would be checked, he told us, by accidents, war, pestilence, famine, infanticide, and murder. Well, it hasn't turned out quite that bad—yet at least. Economic progress, notably developments in food production, has kept most people nourished even if the more intensive use of the world's resources has given rise to scares discussed elsewhere in this book.

Sensible debate around the topic of population growth has been hampered by several factors. One is the population projections, which have a reputation of being fantastically unreliable— they are heavily influenced by the prejudices of those conducting the forecast, and very small differences in assumptions can make large differences to the results, due to the power of compound growth rates. Another problem has been the inability to define overpopulation. Conceptually it can be thought to arise when there is a shortage of resources leading to an impaired quality of life, serious environmental degradation, or long-term shortages of essential goods and services. But how serious is serious and how long-term is long-term? As there is never a eureka moment when we can suddenly say that there is overpopulation, the whole debate is conducted in shades of gray. Population generally changes slowly and unevenly, and the concerns or pressure points vary according to the society's location and its wealth.

Nevertheless, there have been other recent well-known works along similar lines to Malthus. These include *The Limits to Growth*,[2] the world's best-selling environmental book, published in 1972, which modeled the consequences of rapidly growing world population given finite resource supplies, and *The Population Bomb*,[3] which predicted that hundreds of millions of people would starve to death in the 1970s and 1980s. At the time there was no shortage of criticism of the books, and both appeared on lists of the century's worst books made at the turn of the millennium.

Whatever the predictions, it is not difficult to argue that, with around 1 billion people already malnourished and without access to safe drinking water and healthcare, the earth is supporting 6.5 billion people only because many live in misery. Others, more optimistically, have suggested that the world has a "carrying capacity" of nearer 10 billion, and that the falling rate of population growth in various parts of the world, coupled with progress in science and technology, means there will be no problem with overpopulation.

Rapidly growing populations leading to overpopulation might appear to be a global issue, but it is limited to a minority of geographies. Indeed, the United Nations forecasts that nine countries will account for half of the world's projected population increase in the period up to 2050. These are India, Pakistan, Nigeria, Democratic Republic of Congo, Bangladesh, Uganda, the U.S., Ethiopia, and China, listed in order of their contribution to the growth. The growth is associated with increased urbanization, bringing a range of problems, with cities growing particularly fast in China and India—India already has more than 30 cities with populations over 1 million. These changes will cause a shift in where the world's power—in terms of population—is concentrated. All of the growth forecast for the next four decades—adding 2.6 billion to the world's population—is expected to take place in less-developed regions, with the population of the developed world remaining unchanged at around 1.2 billion.

Demographers normally measure either the crude birth rate (the number of children born per thousand of population each year) or the total fertility rate (the average number of children born to each woman over the course of her life). On either measure, a number of developed countries stand out as having a problem with declining population levels. Fertility in several dozen developed countries has reached levels unprecedented in recorded history—below 1.3 children per woman in several southern and eastern European countries. By contrast, fertility at the world level stands at 2.65 children per woman, a figure that rises to 5 children per woman in the least developed countries. The world's average crude birth rate is around 20 children per thousand of population each year, yet a number of developed countries, including Germany, Japan, and Italy, have rates of below 10.

Countries have always feared a declining population. In the past, a large and growing population was required to develop land and generate wealth. Population increase was encouraged, often by means of conquest and enslavement. Larger armies, and increased security, required a healthy supply of youths. The pressures are different these days, but a declining population, often associated with an aging population, is widely expected to damage economic growth and wealth generation, in turn increasing the difficulty of caring for the elderly.

The reasons for the drop in fertility are not entirely clear. The theory of "demographic transition" suggests that, as the standard of living and life expectancy increase, family sizes start to drop. At one level, factors such as the increased access to contraception give

Crude birth rate of selected countries

Rank by state*	Country	Births/1000 population (crude birth rate)	Total fertility rate
1	Niger	50.7	7.5
20	Nigeria	40.4	5.5
55	Pakistan	29.7	4.0
89	India	22.0	2.7
–	World Average	20.1	2.6
112	Israel	18.0	2.4
122	Brazil	16.6	1.9
136	Ireland	14.4	1.9
137	U.S.	14.1	2.1
142	China	13.3	1.7
148	Australia	12.1	1.8
151	France	12.0	1.8
160	UK	10.8	1.7
169	Spain	10.1	1.3
173	Russia	10.0	1.3
174	Poland	9.9	1.2
178	Italy	9.6	1.3
180	Japan	9.4	1.4
181	Singapore	9.3	1.1
184	Czech Republic	9.0	1.2
189	Austria	8.7	1.4
192	Germany	8.2	1.4

Source: The World Factbook, www.cia.gov. *The total fertility rate is the expected number of birth rate.

adults more choice over when and if to have children. The tendency to get married at later ages, reflecting in part the desire of many women to have careers, also reduces the scope to have children. And the sharp fall in infant death rates has reduced one pressure to have multiple births. The financial equation of having children has also altered: In rural areas in less-developed countries, children contribute to the economic well-being often from an early age, but in developed, urban settings bringing up children is increasingly expensive.

The low fertility trends in some countries are such that demographers are now warning of "negative momentum," occurring when a shrinking population goes into an ever-steeper spiral of decline—fewer babies now means fewer mothers in the future. When fertility rates fall below 2.1 (each woman needs to give birth on average to 2.1 babies to maintain a developed nation's population size) and death rates are broadly stable, a country's population will decline unless it is offset by a favorable combination of immigration and emigration.

This is a situation that is now facing much of Europe. The population of many of the former Soviet republics is falling due to emigration (notably since the fall of the Berlin Wall in the early 1990s), ill-health of those who remain, and relative poverty. The population in the Czech Republic, Hungary, Poland, and the Slovak Republic was lower in 2005 than in 2000. The population of Germany declined in 2005 and 2006, a situation that would be reflected in several other western European countries were it not for net immigration. The future looks no more promising for most developed countries. Europe's population was estimated at 731 million in 2005—the base from which the United Nations conducts its projections. On the "low" scenario, Europe's population is expected to fall by over 20 percent to 566 million, by 2050. The "high" scenario sees a rise of 6 percent, while the "medium" scenario sees a fall of 10 percent. All other continents, in contrast, see a significant increase in population in even the low scenario.

A declining population will usually be accompanied by population aging, one of the factors explaining the economic malaise of Japan and Germany in the last decade—and it's going to get worse before it gets better. The median age of the world's population is currently around 28 years, but that ranges from around 16 to 18 in many of the

Population projections, UN's "medium" variant, selected countries

Millions	2005	2015	Percentage change
World	6465.7	7219.4	+12
The Growing Countries			
Brazil	186.4	209.4	+12
China	1315.8	1393	+6
India	1103.4	1260.4	+14
Nigeria	131.5	160.9	+22
Pakistan	157.9	193.4	+22
Turkey	73.2	82.6	+13
U.S.	298.2	325.7	+9
Stable and Shrinking Countries			
Bulgaria	7.7	7.2	-7
France	60.5	62.3	+3
Germany	82.7	82.5	0
Italy	58.1	57.8	-1
Japan	128.1	128	0
Poland	38.5	38.1	-1
Russia	143.2	136.7	-5
UK	59.7	61.4	+3

Souce: World Population Projects, United Nations, www.un.org/esa, 2005 version

less-developed African countries to over 40 in a good number of European countries. Looking ahead, the problems could be severe for Europe's pensioners. Currently there are around 35 pensioners for every hundred people of working age, but by 2050 there could be twice as many pensioners—around 75—for every hundred workers. Italy and Spain could see their ratios approach one for one. In most European countries, pensions are paid out of current tax revenues, which means that taxes will have to rise sharply, with the burden falling on an already proportionately shrinking number of workers, or pensions will have to fall.

There is little doubt that countries take a decline in the population seriously and try to reverse it. Russia's former President Putin has described the baby shortage as the country's most acute problem and ordered parliament to give large financial incentives to women who have a second baby. Australia's "one for Mothers, one for dad, and one for the country" campaign and the associated baby bonus cash payment, introduced in 2004, has tentatively been declared a success with the latest figures showing a small rise in the birth rate. And many European countries, including France, Italy, and Poland, also offer financial incentives to mothers or families with children. But, if the incentives in the form of cash and savings offered in Singapore and Japan over a rather longer period are anything to go by, they are unlikely to have any lasting success. With surveys suggesting that it can cost several hundred thousand euros to bring up a child, it will be surprising if cash sums of €1,000 or €2,000 have any impact on the underlying trends.

One possible solution is to encourage immigration, but that is another story with another set of issues. The poor and relatively youthful countries of North Africa and Asia that are closest to Europe offer a large supply of potential immigrants, but will the aging residents of Europe want them? It seems certain that continued modest migration will play a part in Europe's policy, but the numbers involved are such that immigration cannot feasibly plug the gap left by the birth dearth. Spain would need 170,000 immigrants a year over the next fifty years to maintain a constant population size. Spain's position is extreme, as it currently has few retired people but is forecast to have a large increase. To maintain a constant working age population, it would need an average 260,000 immigrants a year. But to maintain the current potential support ratio (the number of people of working age per older person), Spain would have to accept an annual average of 1.6 million immigrants every year. This is clearly impossible. Europe's predicament might not end in tears, but serious adjustment in policy and expectations is required.

Overpopulation has been a popular theme in fiction, especially in the 1950s and 1960s. Books and films have proposed that the problem can be easily resolved by embracing measures such as raising infants as food or by promoting euthanasia of everyone who reaches a certain age. So it seems paradoxical that the big issue likely to face most readers of this book is the impact of a shrinking population on their wealth and health in their old age.

Family Breakdown

"The U.S. family: 'HELP!'"
—*Time*

The downside of the increasing flexibility that makes later marriage more appealing is that "normal family life" might feel like a thing of the past for most children. The media plays its part, sometimes giving the impression that society is crumbling in the face of delinquent youths, teenage pregnancies, single-parent families, and a high divorce rate. But this panic-inducing story resulting from the rapid change in our society has, in contrast to many others that we live with, a distinctly political flavor.

In the U.S., the two chief political parties hold differing ideologies on who is most reponsible for the proper care of a child. In 1996, First Lady Hillary Clinton famously used the old African proverb "it takes a village to raise a child" during her husband's presidential campaign, stressing the involvement of the entire community when raising children. The Democratic senator-to-be later published a book titled *It Takes a Village*, stating that she chose this title "because it offers a timeless reminder that children will thrive only if their families thrive and if the whole of society cares enough to provide for them."[1]

In 2005, Republican senator Rick Santorum rebutted with a different ideology: "it takes a family." His view holds that "a better America is built one family at a time by strengthening those families and having community organizations and churches there to support the family, as well as the state and federal government and the educational establishment, culture, and news media standing shoulder to shoulder."[2] Indeed, the U.S. government tends to be fairly supportive of the family structure, giving tax breaks to married couples, a tax credit per child in each family, partial coverage of child care expenses, and financial assistance for those who choose to adopt.

And these ideas are not limited to the United States. In the UK, Conservative leader David Cameron has pinpointed absent fathers and family breakdown as two of the root causes of anti-social behavior, and polls show that a majority of the British public agrees with him. He has said that he does not criticize all the single mothers who work hard to give their children a good start in life, but he has pledged to change the tax and benefits system to ensure that there are real incentives for parents to get and stay married.

The British Labour government has taken a different view during the past decade. It says that marriage is "a good thing" but plans no action to support it, saying that policy must be "bias free" when it comes to marriage, adding that love and compassion are what create strong families. While love and compassion are important, it has become increasingly hard, say the government's critics, for families to survive and flourish without the protection of the law and a supportive financial regime. Indeed, traditional families have been hard hit by Labour's tax reforms, including the removal of a tax allowance for married couples and the introduction of welfare credits that reward lone parents at the expense of low-income couples. The structure means that poorer couples, for whom income top-ups from the state are important, are much better off if they keep their relationships unofficial and, as far as the government is concerned, live separately. Getting married means that benefits are cut. One report, "Parents live apart to cash in on benefits system,"[3] suggested that as many as one million British couples in a committed sexual relationship live most of their time at separate addresses, and that such untraditional family structures primarily reflect financial considerations. An unemployed mother who leaves the unemployed father of her children could experience a rise in her standard of living of between 20 and 35 percent.

It is, of course, impossible to be 100 percent sure of any cause and effect when it comes to behaviors in society, but a number of important trends seem to be heading in the wrong direction, damaging the fabric of society in a number of countries. A recent survey found that 25 percent of working dads in the U.S. spend an average of 1 hour a day with their kids. Forty-two percent of the dads surveyed said they spent less than 2 hours a day with their kids.[4] And working moms don't seem to have quite enough hours in the day, either: The 2008 American Time Use Survey (conducted by the Bureau of Labor Statistics) showed that during work days, married mothers employed full-time spent only 1.4 hours a day with their children. This measurement takes into account reading, playing, talking, and other basic physical needs.

Indeed, leaving children to their own devices, letting them learn about life from other teenagers, and not providing good adult role models are frequently seen as being at the heart of the parenting problem. Whatever the truth, the trends seem to be leaving a large number of unhappy and helpless families in their wake. According to a survey in 2008, 71 percent of working moms and 64 percent of working dads feel managing their families is as challenging as—or more challenging than—managing their own careers. This same survey reveals the widespread discontent of working fathers, 55 percent of whom believe that their companies should offer help in achieving a balance between career and personal life.[5] (Incidentally, a survey conducted in three major companies reveals that parents who are concerned about their children's after-school activities tend to be less productive in the workplace.[6])

A comprehensive evaluation of the well-being of children and young people by UNICEF presented a report described as a "damning" assessment of the "crisis of childhood."[7] The report measured well-being across more than twenty developed countries under a variety of headings such as education, relationships, behaviors, and health. The top half of the well-being table was dominated by small northern European countries including the Netherlands, Sweden, Denmark, and Finland. Relative child poverty was highest in three southern European countries (Portugal, Spain, and Italy) and in three Anglophone countries (the U.S., the UK, and Ireland). Roughly 4 out of 5 children in the countries under review lived with both parents, but the rates fell to below 70 percent in the U.S. and the UK. The percentage of children who said that their peers are "kind and helpful" varied from above 80 percent in some countries to less than 50 percent in the Czech Republic, the U.S., and the UK. Nearly one third of American 15-year-olds were reported to have used cannabis, and 11.6 percent of children in the U.S. engaged in activities that put their welfare at risk.

British newspaper reports of the study—"Threadbare family lives" and "Unhappy families" were typical—focused not on Britain's top-of-the-table performance on teenage pregnancies, bullying, and poor family relationships, but on the depressing conclusion that British children perceive themselves to be unhappier than their peers in other countries. Overall, the UK was the country where the welfare of children suffered most, and America was the nearest rival. While politicians, think tanks, and the press could build little consensus about what to do to improve the situation, it is clear that Britain has suffered from adverse trends. The country finds itself in a worse position on a number of counts than many others and is at a loss to know what to do about its delinquent children.

One of the main contributing factors is the growing unpopularity of matrimony in the UK. Marriage rates in England and Wales are at their lowest since records began over 150 years ago. Just 23 men per thousand unmarried men got married in 2006, a drop from a rate of 28 in 2004 and 49 per thousand twenty years before. The sharp drop in recent years was explained by new legislation aimed at reducing the number of "sham marriages"[8] and the increasing popularity of getting married overseas, but there is no mistaking the underlying trend.

In the U.S., marriage has remained at a stable rate (approximately 7.6 marriages per thousand people)[9], but the average age at which women get married is changing dramatically, as women are waiting until later in life to tie the knot (*Bride Magazine* says the average is now 27 years). The pattern of motherhood has also changed dramatically in recent decades even if the number of children born each year is increasing. For over 30 years, the average age of mothers at first birth increased steadily; in 2003 it was 25.2, an all-time high for the nation at the time it was reported.[10]

And this trend is widespread: Mothers with old faces and young children are an increasingly common sight on Europe's high streets as the average age of motherhood continues to rise in every European country for which records exist. The mean age of women at the birth of the first child is now around 27 to 29 years in most developed countries. In Spain, 13 percent of first-time mothers are now over 40. Later births and later marriages are a result both of some women enjoying their freedoms and careers and of some men shying away from the traditional male role of providing financially for a wife and children.

Those women who are lucky enough to fall pregnant in their late thirties or forties get no thanks. As far as many newspapers are concerned, they are simply risking their health and that of the fetus—miscarriages, ectopic pregnancies, and pre-eclampsia are more common among older mothers-to-be. One doctor said that women who delay having children until they are thirty-five or over constitute a "major public health issue," adding that they are more of a burden to society than teenage mothers.

The papers now regularly report women aged over 60 giving birth. The oldest childbearer in the U.S. gave birth in 2007 at the age of 60, but a 66-year-old gave birth in Romania and a 65-year-old in India. This is as much a freak show as anything else but it does show how medical techniques are extending the age at which women can conceive, even though most clinics will not accept women for IVF treatment over the age of 45. Plenty of women who have their first child at an older age end up with just the one child, and the swelling ranks of children growing up without brothers and sisters are giving rise to a new group of people that is of increasing interest to psychologists.

Teenage mothers are a constant source of interest to the newspapers. One story, "U.S. fears of teen pregnancy pact," told of a group of 17 teenage girls in Massachusetts who suspiciously conceived around the same time as one another. And who could forget the (repeated) Jamie Lynn Spears pregnancy saga? Yet the percentage of teen births has actually been down in recent years. In 1990, nearly 43 percent of pregnancies were to women under 25 years old, but in 2004 that number fell to 38 percent. The U.S. birth rate among teens is only half of that seen in a number of eastern European countries, a little higher than that in the UK, and at least double the rate seen in a number of the main European countries including the Netherlands, Germany, France, and Italy.

However, the proportion of births that are occurring outside marriage has increased sharply. In 2004, 6.4 million pregnancies were to unmarried women (that's 45 percent), a significant increase from the 2.7 million pregnancies that were out of wedlock in 1990. (Interestingly, the number of pregnancies among *married* women fell from 4.1 million in 1990 to 3.5 million in 2004.)

Once married, people can divorce—and divorce rates are high in the U.S.; roughly 1 in 2 marriages doesn't last. France, Germany, and the Netherlands have similar rates (between 40 and 50 percent) and the UK's is higher (over 50 percent). The divorce rate in the countries of southern Europe, including Portugal, Greece, Spain, and Italy, is half that of the U.S. They might be seeing strong rises—Italy, for example, has seen the number of divorces rise by 70 percent in the last decade—but they are starting from a lower point. One report suggested that Italy faced a particular problem: The normal explanations for a marriage break-up—marrying too young, squabbling over money, and meeting a new partner—have been surpassed by the problems arising from the unusually close attachment of Italian men to their mothers. Modern, young, Italian women are less readily able, it seems, to offer the type of unconditional love that young men become accustomed to from their mothers.[11]

One academic study did at least conclude that "divorce works," in that a comparison of the mental states two years before marital breakdown with two years afterward showed an improvement, albeit modest, in "mental stress." This is not to suggest, though, that a greater number of couples should dissolve their unions as those who split were normally the less happy among the married population. It seems that an increasing number of divorces among middle-aged couples are being prompted by women in a final bid for personal freedom, escaping from unfulfilling relationships, rather than the stereotype mid-life man dumping his long-term partner for the thrill of younger flesh.

While many children are raised happily and successfully in couples that are not married, the evidence shows unambiguously that married parents are more likely to stay together and much less likely to suffer from low incomes. The children of married parents are less likely to fail at school, turn to drink, drugs, or crime, have mental health problems, or become teenage parents themselves.

In the United States, 1.5 million babies are born to unmarried women annually.[13] This number is slightly higher than in Britain, where 4 out of 10 children are born outside of marriage. One survey suggested that almost half of the children from separated families had not seen their father at all in the last year.[14]

For all the panic about children, the majority of family units across Europe are childless. One in 8 Europeans lives alone and nearly half of European households are heterosexual and single-sex couples living without children. Some have seen their kids leave home while others will never have children, but with increasing life expectancies, these couples are likely to spend the biggest chunk of their life in the company of their partner. There is some survey evidence to suggest that an increasing number of people will now only marry and start families when they are convinced that it will enhance their own lives. This is a massive change from earlier generations, who thought that only a life blessed with children was a fulfilled one.

The rapid societal changes driven by changes in family structure have made the socio-economic classifications, so loved by statisticians, almost obsolete. Dividing people according to the nature of work (managerial, intermediate, semi-routine, unemployed, etc.) of the head of household now looks rather out of date. As a result, several private-sector companies have created their own so-called geodemographic systems for categorizing people and the areas in which they live. Such systems typically have as many as 50 categories.

It seems that societies, individuals, and policy makers are finding it hard to come to terms with the consequences of a rapidly changing definition of "family," which is increasingly flexible with ever more diverse constituent parts. But is it any worse than before? The tip of the iceberg is delinquent children who are drug-taking criminals, forming gangs to terrorize estates, but are the apparently more widespread problems of alcohol, smoking, promiscuity, anti-social behavior, and school attendance having much effect? Solutions are not readily available, but many feel that the government's policies related to social cohesion are going in the wrong direction: People see them chipping away at society's structures when they would rather see them being reinforced and rebuilt.

The Marriage Squeeze

"A good man is harder to find"
—*Newsweek*

Twenty years ago, *Newsweek* magazine published a story, based on Yale and Harvard research, about the difficulties that older women have finding a husband.[1] It brought to the fore the problems of reconciling a career and children and prompted much hand-wringing, countless newspaper articles and TV debates, and no doubt provoked many a tearful discussion between single women and their mothers.

It started:

> The dire statistics contained in a new demographic study confirm what everybody has suspected all along: many women who seem to have it all will never have husbands. White, college-educated baby boomers in particular are victims of a marriage squeeze—a shortage of available men that adds up to a numbers game that women can't win.

It went on to say that these white, college-educated women "born in the mid-1950s who are still single at 30 have only a 20 percent chance of marrying. By the age of 35 the odds drop to 5 percent. Forty-year-olds are more likely to be killed by a terrorist: they have a minuscule 2.6 percent probability of tying the knot." The catchy reference to terrorists was, it seems, journalistic license and did not appear in the academic report.

The story sparked a crisis of confidence among America's growing ranks of single women, and the core message quickly became entrenched in pop culture, even though the study was widely criticized. The media was blamed for inventing "a national marital crisis on the basis of a single academic experiment of dubious statistical merit."[2] Some saw it as a backlash against feminism, with one columnist writing that the report gleefully warned women: "Reach too high, young lady, and you'll end up in the stratosphere of slim pickings."

The U.S. Census Bureau responded by publishing far more promising probabilities, suggesting that 30-year-olds had a 60 percent likelihood of finding a husband, with the odds dropping to around 20 percent for 40-year-olds, respectively three and eight times greater than the odds in the original article.[3] With the benefit of hindsight—two decades on we can see what actually happened to these women—and looking at the personal experiences of the two authors of this book, the notion that "women are over the hill at 30" looks a little bit archaic. The latest U.S. census records show that only 1 in 10 college-educated women now aged in their fifties has never married.

The mistake in the research was apparently a simple one— namely to assume that current and future generations would behave as past generations did. Just because, for example, most women used to be married by 30 in the past and those that did not tended to remain single, did not mean that those not married by 30 in the future would also remain single. The *Newsweek* episode offers a cautionary tale of what can happen when the media oversimplifies a complicated and less than perfect piece of academic work. Nonetheless it clearly hit a nerve as it did in part reflect the experiences of many people.

These fears and worries have been dramatized in recent years in humorous blockbusters which ensure that the marriage-prospects blues remain once the laughter has died down. *Bridget Jones's Diary* told the story of a single woman over 30 who smokes too much, drinks too much, and has a tendency to say whatever comes into her mind. Her mother keeps setting her up with dorks, and she has an awkward fling with her boss, leaving us wondering if she will ever find a husband. *Sex and the City* told of the antics of four attractive female New Yorkers who gossiped about their sex lives—or lack thereof—and searched for new ways to deal with being a woman as they navigated the turn of the millennium.

So, after the magazine- and big-screen-drama-induced panic, what are the facts? In most western countries, there are roughly equal numbers of men and women at the crucial ages. Although more boys than girls are born—roughly a ratio of 105 to 100—male teenagers and men have higher death rates than their female counterparts, prompting the numbers gap between the two sexes to narrow until some point in middle age when women start to outnumber men.

In the UK, the official population estimates are very evenly balanced between the two sexes from the ages 20 to 44, but the reality is probably more favorable than that for women, as it is widely believed that the population census in 2001, on which these figures are based, undercounted the number of young to middle-aged men by several hundred thousand. An even more favorable situation for women prevails in the U.S.: There are 2.2 million males age 25 compared to 2.1 million females, and it is only at age 42 that females start to outnumber males.

If there are enough men, the root of the problem is changing attitudes and behavior patterns. Marriage has been hit hard by the increasing tendency to cohabit, but unfortunately there is very little decent data telling us exactly how relationships are evolving over time. People are behaving more flexibly—moving in and out of relationships, including marriage, more frequently than in the past, making it harder to track and monitor. And the definitions are complex in any case—would people in long-term relationships but living separately, an emerging trend especially among middle-aged people with established lives, be said to be single? And what about the increasing number of people who are living an openly gay or bisexual life?

What is clear is that marriage has been pushed later with each decade for all post-war generations. In the year 1900, American males married at an average age of 25.9 years and the average woman who got married was 22. These days, the numbers are higher: 26.8 for men and 25.1 for women.

The explanations for delaying marriage and childbirth are many and varied. Increased graduation rates and changing career patterns have certainly played a part. Hard-working professional people are often expected to work hardest in their twenties and thirties, making it difficult—especially for women—to find time to marry and have children. Many people simply enjoy being single at a time when it has become acceptable, in a way that was not the case a generation ago, and set up home on their own. Indeed, one survey suggested that 97 percent of British people aged between 25 and 34 believed that it was important to live alone before settling down.[4] One rather depressing survey suggested that people do not enter relationships because they don't expect them to last.[5] And a new "social class," dubbed the "regretful loner," has been coined for men in their late thirties and forties who live alone and have either failed to form relationships or are the victims of failed relationships.

But for most people, being single is just a phase. If surveys are to be believed, most women say that they want to get married, but at the same time they seem to be asking themselves about the cost in terms of scaling back their careers or sacrificing their (probably) enjoyable single lives. Men, too, it seems, are also keen to marry—as indeed they ought to be, as surveys suggest that married men are healthier and happier than their single counterparts. But, according to the media, "the problem" of falling in love with someone suitable to marry is more of an issue for women. It is certainly true that nearly all newspaper articles on the issue are written by women. Whatever the truth, much of the media coverage leaves us with the impression that the men who don't marry come from the bottom of the barrel while the women who don't marry are the cream of the crop.

While the reasons for delaying cohabitation and marriage are perfectly sensible, it brings a price. Surveys have shown that single people—male or female—consume more alcohol, often work longer hours, tend to skip meals and eat less healthily, and might well not have someone close to share life's problems with.[6] It is also more expensive to live on your own. Most notable, however, is the anxiety related to the biological clock and that many women still desire to have children. In essence, the more fun or the more career that a woman has, the less time she leaves to find a husband and to produce babies.

As ever, when faced with a challenge, behaviors change. Women in their thirties have set about the task of finding a mate with the same efficiency that they might bring to their professional careers. The booming singles industry, based around the Internet and variants of speed dating, has emerged to cater for all needs. And artificial insemination is now widely available if the priority is to have a child.

In most western countries, the situation is not too bad: it boils down to difficult personal choices in rapidly changing social environments, and of course, a good dollop of luck. As one woman said, "I'm not a statistic. I am one woman. I need to find one man." Most people do eventually marry, if they want, despite the demographic forecasts. And most manage to juggle careers and children without screwing up too badly. It is also said that hooking up with someone later in life is less of a gamble than doing so earlier on—certainly the divorce rates are lower—as both parties have matured and have a better idea what they want from life.

Whatever the problems in the West, relationship-forming is likely to be much more problematic for the young in China, where a severely unbalanced sex ratio will leave millions of men unable to find brides. Sons have been traditionally preferred in China, and most couples can have only one child, so many prospective parents have, over the years, aborted pregnancies if tests showed that the fetus was female. As a result, there are 119 boys for every 100 girls in China, well above a global ratio of no more than 105. In some parts of China, where the ratio has been even more distorted—as high as 134 boys to every 100 girls—programs have been launched, largely successfully, providing benefits and cash payments to encourage families to have girls. Meanwhile, the State Commission for Population and Family Planning estimates that there will be 25 million men who will fail to have wives by 2020. How long before a male Bridget Jones comes out of China?

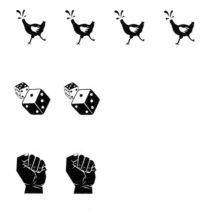

Something for the Weekend?

"Sex with many partners? No thanks, we're British"
—*The Times (of London)*

Virtually all western homes have a toilet, a sink, and a bath or a shower. But one item of bathroom equipment shows that hygiene habits are not universally shared. It is the bidet: "This hygienic French invention of the 18th century has taken the world by storm," according to the French journalist Agnès Poirier writing in Britain's *Observer*. "Only the USA and Britain are bidet-illiterate."

Market evidence confirms the truth of this. Multinational sanitary-ware manufacturers find they sell very few bidets in the United States, whereas in Italy they sell 71 for every hundred toilets.[1]

A dictionary definition describes the bidet as an installation for "bathing the external genitals and the posterior parts of the body"—not, as some travelers believe, a receptacle for washing socks. So, in countries where bidets are prevalent, do people wash their genitals more often? And, more to the point, do they do it because they are getting more sex, as some Americans suspect? Or is using a bidet just a substitute for a proper wash, something Americans also suspect, mindful as they are that the French buy less soap than they do and that Napoleon preferred Josephine not to wash before they made love. In which case they are probably still getting more sex, the dirty beasts.

"More Sex Please, We're French" is typical of the headlines that fuel American anxiety that they are missing out compared to Latin lovers. Such anxieties are helped along by surveys like the one carried out in 2002 by the condom manufacturer Durex in which French people claimed to have sex 167 times a year, the British 149 times a year, and Americans just 138 times.

The Durex survey has grown each time it has been done. In 2005, it polled 317,000 respondents in 41 countries. This colossal undertaking sounds authoritative, but it is not. Researchers into sexual behavior and attitudes criticize the survey's leading questions, its biased sample of respondents limited to those with Internet access who feel like filling in a questionnaire, and its apparent lack of ethical safeguards.

Its shortcomings are revealed when you try to compare the results over time—something not encouraged by Durex, which updates the results on its website and not in academic journals. Recall that in 2002 the French were claiming to have sex 167 times and the Americans 138 times a year. But by 2005, the figures were 120 and 113.2, respectively. What made everybody so frisky in 2002? The Chinese went from being the most faithful lovers to being the most promiscuous in the space of two years, according to the same survey. Hardly likely.

The inability to compare results from one year to the next doesn't bother the media, however. Durex's context-free findings guarantee a titillating headline. One country will always have the sexiest girls, the highest level of unprotected sex, or the most one-night stands—and if it's a different one each time then so much the better.

This kind of survey, which makes it easy to compare regions or countries (but not different time periods, age groups, or social classes), encourages us to regard sex as something that is appropriate to put in league tables, like hospital waiting lists or school exam results. This is clearly nonsense since sex is a matter for individuals and is not administrated in this way. Whereas our grandparents supposedly grew up fearful about sex because of widespread ignorance, today we grow up fearful about sex because we are not having as much, in as many ways, with as many people, as everybody else. The recent addition to the Durex survey of a number of Asian countries, for example, reveals a greater sexual conservatism among these respondents compared to their peers in the West. Presumably the company hopes that their nations' "low placing" in the tables may lead some young Indians and Indonesians to become more sexually active—thereby providing new market opportunities.

Many commercial interests have now seen how easy it is to gain publicity by adding the spurious authority of a "survey" to the sure-fire ingredient "sex." Durex is merely the leader in an increasingly crowded field. "Much more sex please . . . we're British" was the headline to news of a poll by a New York marketing agency.[3] This pleased the media because it confounded the usual stereotype. "Single Britons are the most promiscuous in the world, an international survey of sexual attitudes says," the *Guardian* reported enthusiastically. An entirely different slant could have been put on the same data. The poll showed, for example, that two-thirds of British men felt entitled to regular sex with their partners, but that less than half of women did. In the four other countries surveyed, these figures were considerably higher for both sexes. One could equally have concluded that the British are closer to giving up altogether on regular sex with their partners than the French, Germans, Americans, or Chinese.

There again, perhaps the Anglo-Saxons have the last laugh. The myth of the Latin lover went limp recently when the *New York Times* headlined "Spain says adiós siesta and hola Viagra." The story concerned a man who held up a Madrid pharmacy, demanding the erectile dysfunction drug Viagra. The article went on to theorize that the exploding Spanish market for such drugs was due to the abandonment of the siesta and the discovery of work ethic, leaving men too stressed to perform in bed.

Occasionally, the message of constant sex is contradicted by these polls. In hopes to fight the misconception that everyone is doing it, the Department of Health and Human Services launched a campaign to the contrary, supported by statistics that show that 53 percent of high school students have not had sexual intercourse. The "What's The Rush" campaign launched in Tennessee refers to statistics in Ontario showing that 55.1 percent of males and 62.5 percent of females under 16 have not had sexual intercourse.[4] Interestingly, Durex's separate global survey produced results not incompatible with this—giving 16.9 as the age at which most American people lost their virginity, compared to around 18 in Latin countries. But on its website, Durex ignored the newspaper's take on the national result and, with an eye for the market, drew attention instead to its global result that "the trend is for people to lose their virginity earlier."

One newspaper at least seemed to have the measure of things. Under the headline "Teen sex surveys ruin your love life," the *Daily Telegraph* lampooned the tendency for these heavily publicized polls to highlight promiscuity among the young: "Many young people claim to be regularly taking part in two surveys a day, often with different research organizations."

Newspaper readers are not all young, of course. What about something for the mature reader? In 2007, the Associated Press reported that "Sexed up seniors do it more than you'd think." According to the article, "An unprecedented study of sex and seniors finds that many older people are surprisingly frisky—willing to do, and talk about, intimate acts that would make their grandchildren blush." A similar survey, the portentous-sounding Global Study of Sexual Attitudes and Behaviors drew 30,000 respondents. It unaccountably failed to mention that it was sponsored by Pfizer, the maker of Viagra. Its target market may be older, but Pfizer's message is essentially the same as Durex's—you too should be getting more, and our products can help you get it. Both companies know that the best way to suggest this is to insinuate that your peers are already at it like rabbits.

Commercially motivated surveys may be unreliable, but, when it comes to sex, it is more frequently observed that those surveyed might be the ones being economical with the truth. This possibility affects any survey, but the suspicion tends to be raised most often with sex surveys. Men often exaggerate their experience whereas women downplay theirs. Americans might expect Latin lovers to brag about their conquests. There is no way of knowing unless an expert closely questions the respondents directly, but then the pollsters do not need to have the true picture in order to generate media interest.

All this represents something of a wasted opportunity. Data on sexual attitudes, properly gathered and analyzed, could inform social and healthcare policies rather than merely satisfying our prurience. They might even have something useful to say about the divergence of birth rates in different countries. It is one thing to have young survey respondents boasting of their promiscuity, for example, but establishing how this behavior—and underlying attitudes—relates to sexually transmitted infection and teenage pregnancies requires expert collection and analysis of the data. For fear that it will offend sponsors or advertisers, no popular survey is likely to have much to say about these or other important topics where little is known, such as the extent of sexual ignorance, harmful practices, and violence related to sex.

Academic surveys that do tackle these trickier issues have their own troubled history. Alfred Kinsey's 1948 report "Sexual Behavior in the Human Male" opened the floodgates for sex research. But Kinsey's follow-up study of women in 1953, and a British survey by the social anthropology project known as Mass Observation in 1949, which included both sexes, were both regarded as shocking because they reported a high proportion of women having pre-marital sex and affairs once married. The latter study did not see the light of day until 56 years later, when the BBC promptly dubbed it "Britain's secret sex survey." One in 4 men said they had had sex with a prostitute. One in 5 women admitted to having affairs outside marriage, and more than half of both men and women had had sex before marriage. Homosexual activity, then illegal, was also reported by 20 percent of both sexes.

Trying to find out what really goes on in bedrooms remains controversial today. In 1990, a National Survey of Sexual Attitudes and Lifestyles in Britain by epidemiologists and social researchers at the University of London was reported to have been banned by Margaret Thatcher's Conservative government; it was eventually funded by the Wellcome Trust and published in the usual way. The survey was repeated in 2000 with less attendant fuss. As well as providing important information about the prevalence of HIV, AIDS, and other sexually transmitted infections, the surveys showed an increase in various promiscuous behaviors over the decade. Nevertheless, by 2000, more people were using condoms, including on their first time of sexual intercourse, and the number of girls under 16 having sex had stabilized. The 2000 report also found, in marked contrast to the Mass Observation findings, that only 2.6 percent of respondents reported homosexual experiences while 4.3 percent of men had paid for sex.[5]

The British media gave the findings the serious consideration they give to any sex survey, combining moral indignation with facetiousness. The *Mail on Sunday* reported that "one in 20 married men had been unfaithful in the past year, while only one in 50 married women had strayed," using the figures to confirm "the received wisdom that men commit much more adultery"—which is not quite what they say. The tone of the words "unfaithful," "strayed," "commit," and "adultery" is to be contrasted with the researchers' more neutrally phrased questions about whether married persons had taken "new partners."

Such national surveys can now be put together to build an international picture of sexual activity—a picture that turns out rather different from the debauch painted by Durex. In 2001, Kaye Wellings, one of the authors of the British survey, and colleagues in France, South Africa, and the United States gathered survey data for 59 countries.[6] Unlike Durex, they did not ask about threesomes, bondage, or sex on the beach. They found, boringly, that there was no global trend toward earlier sexual intercourse and that married people have the most sex. There was a global increase in premarital sex—mainly because people are getting married later. Unlike the heavily marketed Durex survey, this global study prompted only one newspaper story, under the drearily predictable headline "Sex with many partners? No thanks, we're British."

Some curiosities might have aroused more media interest. Promiscuity was generally increasing in developed countries, nowhere more than among Australian women, who have caught up and are overtaking men in making full use of their single years. Everywhere, men reported more multiple partners than women, but in Brazil the disparity was so great that the researchers could only explain the result by men's over-reporting of their prowess due to the "Latin macho culture."

As these contrasting examples show, you learn what you want to learn from sex surveys. Academics have, in Wellings's words, "a historically unique opportunity to describe patterns of sexual behavior and their implications for attempts to protect sexual health at the beginning of the 21st century." For Durex, the priority is to represent sex as a recreational activity. The beautiful twenty-somethings beaming smiles from its website and telling us about their favorite sex enhancers make that clear enough. For Pfizer and other pharmaceutical companies, sex is a huge new market opportunity. But for this market to come about we must see sex not as a mystery or as fun but as a medical problem that can be solved with the aid of drugs.

The pharmaceutical industry has new plans for drugs to treat premature ejaculation, heighten the intensity of orgasms, or simply make you want more sex. Though ostensibly aimed at bringing help to a minority of genuine sufferers from sexual dysfunction, these seem destined to follow the path taken by Viagra, further encouraging the public to regard sex as a problem for which drugs are the cure. If this happens, the sex surveys and the media's giggling uncritical attitude toward them will have played a large part in getting us there.

2. Health

Unless confronted by immediate danger, we generally regard our health as our prime concern. Our fear of illness is exploited by governments, who want to protect us in the most cost-effective ways, by companies who want us to pop their pills, and by food manufacturers who would like us to believe that their provender is health-giving. But whether or not we worry about our health, we still die. It's worth remembering that we do so on average at a greater age than in the past, with less suffering and fewer declining years.

The Fat Thing

"Overweight and obese kids: Are enough being diagnosed?"
—*Los Angeles Times*

Obesity is a favorite panic story—the media loves everything to do with the subject. Even responsible newspaper editors find stories on the subject as resistibly as a child finds a chocolate bar. The story might be serious—obesity shortens life and costs national health systems substantial funds—but there is also a lighter side—stories suggesting that we will all soon be too fat to fit into airline seats and too heavy to be carried to our graves on pallbearers' shoulders.

"The fattest children in the world" was the attention-grabbing headline of an article warning Britons that one-third of Scottish 12-year-olds are overweight and that one-fifth are obese. The kids are fueled by soda and unfamiliar with fresh fruit, it seems. The Scottish rate of child obesity is above that in the U.S. at 16 percent, and much higher than in other European countries (for example, Ireland 9 percent, Spain 9 percent, France 4 percent, Sweden 5 percent, Denmark 2 percent). A spokesman for the International

Obesity Taskforce was quoted as saying, "The obesity epidemic is escalating totally out of control in Scotland," adding that this "is more than just a warning signal, it's a red light."

The terms "overweight" and "obese" are generally used loosely but they do have a statistical definition, relating height to weight. A person who is 5 feet 9 inches tall would be categorized as overweight at 169 pounds and obese if they weighed over 203 pounds. The thresholds are sometimes expressed in terms of a body mass index (BMI, or Quetelet index, after the Belgian who established the measure 170 years ago), an indirect measure of the amount of body fat. It is calculated by taking the bodyweight in pounds and dividing it by the height in inches, squared, and then multiplying the result by 703.

Someone is overweight with a BMI of over 25 and obese with a measure of over 30.[1] At the extremes, a measure of over 40 is described as morbidly obese, while under 18.5 is thought to be underweight and might well indicate malnutrition or an eating disorder. As the measure becomes more widely used, with doctors using it for medical diagnosis, so the controversy surrounding the measure's accuracy in relation to levels of body fat has increased—it can be distorted by factors such as fitness level, muscle mass, bone structure, gender, and ethnicity. Yet for all its faults, for most people it gives a broadly accurate assessment and it is often the measure used to create data.

The facts behind the bold headlines are scary. A few Arab countries and some Pacific islands have exceptionally high rates of obesity, but among developed countries it is the U.S. that tops the table as obesity among adults has risen significantly in the U.S. during the past twenty years. Nearly one-third of American adults—over 60 million people—are obese, and the proportion of young people that are obese has more than trebled since 1980. U.S. authorities want to halve the prevalence of adult obesity by 2010, but the situation continues to worsen. Louisiana, Mississippi, and West Virginia might be the fattest states, but only 4 have an obesity prevalence rate of 20 percent or less.

Adult overweight and obesity rates, selected countries from around the world, percentages

	Male		Female	
	Overweight	Obese	Overweight	Obese
Australia	48	19	30	22
China	30	2	29	6
Czech Republic	49	25	31	26
England	44	23	35	24
Germany	53	23	36	23
India	4	1	4	1
Ireland	46	20	33	16
Italy	42	9	26	9
Japan	24	3	17	3
Russia	31	10	27	22
Saudi Arabia	42	26	32	44
Scotland	43	22	34	26
Tonga	37	47	23	70
U.S.	40	31	29	33

Source: www.iotf.org. Age ranges covered vary between countries. Data are latest available in October 2007 and mostly relate to a year between 1999 and 2003. Some countries, such as Canada and France, are excluded as their surveys are "self-reporting."

The obesity plague has not visited every developed country but it has gone global—the developing world is suffering as a result of changes in diet, physical activity, health, urbanization, and nutrition, with the bitter irony that, while some developing countries continue to focus their efforts on reducing hunger, others face the problem of obesity. As poor countries become more prosperous, they acquire some of the problems along with some of the benefits of industrialized nations. A hundred years ago just 10 percent of the world's population inhabited cities. Today, that figure is around 50 percent. Cities offer a greater range of food choices, generally at lower prices, and urban work often demands less physical exertion than rural work. Traditional diets featuring grains and vegetables are giving way to mass-produced meals high in fat and sugar. Being overweight used to be a sign of wealth, but now it often marks poverty. The number of overweight individuals worldwide now rivals those who are underweight.

The increasing rates raise concern because of their implications for health. As one newspaper put it, the weight gain "has been so fast and so extreme that experts believe these children will suffer a lifetime of horrific and crippling health problems." Being overweight or obese increases the risk of many diseases and health conditions, including hypertension, type 2 diabetes, coronary heart disease, stroke, gallbladder disease, osteoarthritis, sleep and respiratory problems, and some cancers.[2] It can also lead to inactivity and mood disorders.[3] According to the World Health Organization, a good part of the heart, cancer, and mental-disorder disease burden for the world, including that of most low-income countries, is rooted in the intake of excess sugar, fat, and salt, and a paucity of fruit and vegetables. Obesity is proving to be a "remarkable amplifier" of diabetes, high blood pressure, and high cholesterol levels particularly among Asian and Central American communities, it said.

Obesity is also linked to early death—although the closeness of the link is unclear. The uncertainties were highlighted in 2005, when two studies from the Centers for Disease Control and Prevention produced two very different estimates for the number of deaths due to obesity—1,000 deaths a day or only 26,000 a year. As being overweight does not kill you directly, the way a heart attack might, we will always be left with a—probably wide—range for the number of deaths caused by obesity. Even so, the message would seem to be clear. The chairman of the National Obesity Forum in the UK said that obese children are twice as likely to die by the age of 50,[4] while a Canadian paper explained that today's children will be the first generation of kids who are not going to live as long as their parents. The World Health Organization estimates that more than 1 in 10 deaths in developed countries is already due to overweight and obesity.

The "public health time bomb" of obesity will bring a "vast" cost to national health services around the world, in addition to the direct financial and economic consequences faced by afflicted individuals. One study suggested that the medical expenses associated with overweight and obesity amounted to over 9 percent of the total U.S. medical expenditures—and that was a decade ago, when the problem was less pressing. Estimates of the direct costs are only illustrative in most countries, and up-to-date figures simply do not exist. Most such analyses acknowledge that published figures are likely to be underestimates.

Measuring indirect costs is even more difficult, but they are widely thought to be several times higher. If the World Health Organization's estimate of early deaths is broadly correct, it would imply the loss of several tens of thousands of years of working life in the U.S. and associated earnings, in addition to the cost of disability benefits paid to those unable to work. Official reports estimate that obese employees are more than twice as likely to experience absenteeism: seven or more absences due to illness in a short period of time.

Despite the very real nature of the problem, policy initiatives around the globe are modest and of limited success, reflecting in part that the explanations for the trend are complex and uncertain. Generally higher obesity levels are put down to a rising consumption of fast or unhealthy food and an increasingly sedentary lifestyle in front of television and computer screens. But there is no shortage of alternative explanations paraded in the media—obesity has been linked to a virus, meaning you can "catch fat" from other people, too little sleep, and eating just one plain biscuit a day. One article says, "The obesity epidemic in our country has spared no age-group, even our very youngest children," reporting the rise in the number of children under 6 classified as overweight. And, of course, poor reporting can confuse. The lack of policy progress could also be due to excess political correctness and the nervousness of people across society to speak openly about weight issues with fat people. Some doctors apparently shy away from discussing weight with children for fear of hurting their feelings.

The policies being followed include food-labeling schemes, encouraging food manufacturers to reduce the amount of salt, sugar, and fat in pre-prepared meals, and healthy eating advice, along with a raft of measures for schools, including improved school lunches, changes to the way children play, increased sessions of sport, and the reintroduction of home economics classes, with the curriculum being used to reinforce messages about healthy eating. In 2005, one of America's favorite puppets was put on a moderation diet; for the now somewhat questionably dubbed Cookie Monster, fruits and veggies are a dietary staple and cookies are only a once-in-a-while snack. The hope is that children will take cues from his eating habits. In the UK and Australia, television "junk food" advertisements are being progressively banned for programs with a significant proportion of younger viewers.

But in the face of poor progress, other more desperate measures are being dreamed up: higher taxes and insurance premiums for overweight people; a restriction on government-subsidized health care for people who refuse to lose weight; new building codes that would force people to walk further as they go about their daily lives; and the use of taxes and subsidies on bad and good foods to encourage healthy eating patterns. It has even been suggested that the advertising of weight-loss dieting fads should be banished on the grounds that they give unfounded hopes and obscure the relatively simple messages that public health professionals would like to convey.

Despite all the evidence, some people believe that the "war on obesity" is over-hyped by government and health professionals, and is one of the clearest examples of an ever more intrusive "nanny state." Obesity myth articles typically set out a string of "facts," claiming that, among other things, nobody seems to know why the rise in obesity is happening, we are eating less fat than in the past, some statistics are "pure fantasy," and a majority of children are exceeding the recommended daily physical activity levels. They also pour scorn on the so-called healthy lifestyles solution to the problem, suggesting that badgering large numbers of people to change their lifestyles is a lost cause.

The food industry itself generally recognizes that obesity is a problem, but manufacturers consider it to be their role to provide food that the public "wants," believing that eating is an individual responsibility. As the Canadian Council of Grocery Distributors has said, no doubt speaking for many trade associations, "supermarkets are not in the health regulation business." Although it is fashionable to attack the retailers, many slim people probably wonder in quiet moments why it is that more fat people do not address their weight status themselves.

While some people are becoming fatter, others are eating far more healthily than at any time, benefiting from a greater selection of fresh produce in the shops than ever before and the availability of cookery books and programs and engaging in enjoyable and structured exercise. We are witnessing the emergence of a health gap in our society to rival the existing divide caused by money.

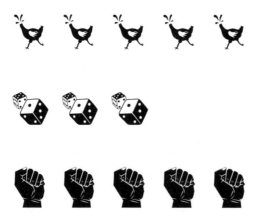

Currying Flavor

"For health's sake, curb the taste for salt"
—*The New York Times*

"Can flavorless food be eaten without salt?" Job complains to God in the Bible. It's a rhetorical question, but it goes to the heart of our difficulty with this essential commodity today. Sugar, fat, and additives in junk food lead you to obesity, but not if the salt gets you first. For salt is now the "hidden killer lurking in your favorite Chinese and Indian takeaway meals," according to the British tabloid *Sun* in an article headlined "Is your curry killing you?"[1]

The headline was prompted by tests conducted by trading standards offices. More than half of 50 Indian and Chinese meals tested contained more than the *daily* adult intake of six grams of salt recommended by the government. One dish of chicken with cashew nuts in yellow bean sauce contained 15.8 grams of salt.

Salt is a simple chemical, sodium chloride, and is the body's main source of these two elements. However, excess sodium intake has been linked to high blood pressure, a major cause of strokes and heart disease. The *Sun* is thus able to claim that "salt-related deaths" are running at 35,000 a year in the UK, though large-scale scientific studies of diet and blood pressure find it hard to isolate salt as the decisive factor.

To clear up the confusion about salt, we need to look at why and how it finds its way into what we eat. We have used salt for millennia not only to enhance the flavor of food but also to preserve it. Chefs therefore prize salt for the same reason as the producers of junk food.

For food lovers such as Hervé This, a chemist at the National Institute of Agronomic Research in Paris and one of the founders of the culinary movement of molecular gastronomy, there are "two great fears: gout and a diet without salt."[2] Salt, he explains, "increases the ionic strength of aqueous solutions, making it easier for odorant molecules to separate themselves from food." In other words, it is a vehicle for bringing out other tastes. This is the case whether the other tastes are subtle and delicious, as they may be in a velouté of asparagus, or faint and banal, as in a processed cheese. It may draw out elusive and exotic natural flavors or it may simply mask skimping on costly flavorsome ingredients.

Unfortunately, the high solubility of salt in water which enables it to release food flavors is also the property that effectively disguises its presence in many foods. We enjoy snack foods because their saltiness is one flavor we seek when we choose them, but then we are surprised and maybe shocked to find that corn flakes, say, contain as much salt as potato crisps because we do not think of them as salty.

It is this disguised presence of salt in western diets that has become a major issue. Up to 80 percent of the salt we eat is contained in processed and packaged foods such as bread and meats. A spokeswoman for the American Dietetic Association advises consumers to "take stock of the sources of salt in your diet, such as restaurant meals, salt-based condiments, and convenience foods. Some of these are really loaded with salt."[3] So it's not that we are choosing directly to put salt on our food.

In some manufactured products a small amount of salt is almost essential. Bread and bakery items usually include a little salt for flavor. But there are more pernicious reasons for adding salt where it is not strictly needed. Salt enables some foods to hold more water, a phenomenon exploited by manufacturers cheaply to bulk up the weight of their product. A supermarket brand of bacon can contain sufficient salt that you would consume your entire daily allowance in no more than three rashers, whereas bacon from a "gourmet" supplier might contain a quarter of the amount.

So how much salt do we eat? In the 1980s, before it was widely known to be associated with high blood pressure, salt consumption in the United States was between 6 and 15 grams a day, according to food scientist Harold McGee, "a dosage that probably supplies 5, 10, even 25 times as much sodium as we actually need."[4] What we need, what we want, and what the food industry wants to feed us are very different things, however. The WHO target daily intake is 5 grams, but national governments are happy to sanction higher levels, which are reprinted on many food packets. But we still eat more salt than this. On its website, the European Salt Producers' Association proudly—if perhaps a little incautiously—touts a figure of 8 grams a day per capita salt consumption. Americans still consume around 10 grams a day.

The producers are vigorous in their defense of people's right to consume as much salt as they want, in tones that at times recall the tobacco lobby. There is no need for healthy people to reduce their salt intake, they insist, while casting doubt on studies linking sodium to high blood pressure. In some cases, they point out, elderly people have died apparently because they have not been getting enough salt. Although the 6 gram daily allowance applies to adults of all ages, the elderly are more susceptible to high blood pressure and so presumably more likely to act on heightened fears by cutting out salt. Not all people should automatically reduce their salt intake, therefore.

But salt is not like smoking, because you aren't always aware of it when you indulge. The recommended daily allowance is well publicized, but this information is of little use if you cannot calculate your intake. This is almost impossible to do. Packaged foods have long been obliged to list their major ingredients, which often include salt, but they do not have to declare the relative amount of salt present. More recently, in response to concerns not only about salt, but also about fats and sugar, manufacturers have begun to include panels of "nutrition information," and some also give overall "guideline daily amounts" of these dietary elements. In the UK, this apparently helpful gesture has been viewed as a preemptive measure to head off a "traffic lights" scheme proposed in 2005 by the Food Standards Agency to display much more readily understood red, yellow, or green gradings for these substances.

But even declaring salt content is not transparently done. Some global brands such as Heinz and Kellogg's responsibly give figures for salt and for that salt in terms of its sodium content alone. Cereals are especially assiduous about displaying this information, perhaps because it is at breakfast that we are most likely to pause to consider our dietary health. But many products indicate salt *only* as sodium. In a sense, this is medically useful since sodium is the component of salt linked to high blood pressure. But it is helpful to the manufacturers too, as 5 grams of salt, for example, corresponds to just 2 grams of sodium, which makes the danger appear less to consumers not fully briefed on the chemistry. In fact, although sodium and salt can be shown interchangeably on food labels, they are not necessarily equivalent at all, as other ingredients such as baking powder also contain sodium.

Unsurprisingly, it is products high in salt that prefer to focus on sodium content. There are 5.75 grams of sodium in 1 tablespoon of soy sauce, for example. This is equivalent to 14.5 grams of salt, well on the way to being a saturated solution of the stuff. A single cube of chicken stock may contain more than 4 grams of salt, although the information may be given in less readily interpreted form as the amount of sodium in a (small) portion of made-up stock.

Newspapers have latched on to the problem, though their headlines can rather miss the point. "Cheese saltier than ocean" was the British *Independent's* line when anti-salt campaigners looked at the label on Kraft's Dairylea Light cheese slices. The cheese does indeed contain more salt weight for weight than seawater—a level boosted by having selected for testing the "Light" variety of the cheese, which contains proportionately more salt simply because it has proportionately less fat. Its more fattening cousin brand is a little less salty than the ocean—which of course makes for a less satisfactory headline. The reader is asked to recall how disgustingly salty a mouthful of seawater tastes and to draw the inference that the cheese tastes equally salty. However, the salt in seawater is the only thing that gives it a taste, whereas the salt in the cheese is used to bring out other flavors, and the overall taste is not overwhelmingly of salt. But, whether you realize it or not, the salt is really there—2.8 grams of it in every 100 grams.

High levels of salt can lurk within food precisely because it mingles with other flavors. This becomes a greater problem when there is no labeling. Highly flavored dishes can hide more salt than bland food. Yet if they are bought as take-out food they are unlabeled for salt or any of their ingredients. In cases where the take-out food is more standardized, this situation may be about to change. McDonald's has begun placing codes on some of its packaging that, when scanned, will transmit nutritional information to the consumer's cell phone.[5]

But this elaborate routine merely makes the point of food campaigners such as Joanna Blythman, who identify an urge to see food not as a pleasure but as "a complex problem" for which eager manufacturers supply "meal solutions."[6] European products tend not to display panels of nutrition information and so give less data about salt, but their consumers tend to be more knowledgeable about food and are more likely to cook from scratch in any case.

Blythman has an unlikely ally in the food producers, who agree that debate about the levels at which the food regulators set their salt reduction targets "misses the point about how healthier eating habits can be achieved."[7] They argue that it is pointless to set "arbitrary targets, which aren't even enforceable by law" when it is consumer power, and not regulation, that is demanding lower salt levels. Manufacturers might prefer to see no targets. But it is clear from the way that salt is covered by a media keen to titillate as well as to inform a largely ignorant public that the guideline daily amounts, "arbitrary" though they may be, serve a vital purpose when presented alongside declared or tested levels of salt in alerting consumers to potential dangers to health.

Where does this leave the health-conscious gourmet with a craving for salt? For these people there is a whole different marketing game. Most things you buy in a grocery are more or less complex mixtures of basic ingredients. But salt is one of the simplest chemicals there is. Salt is sodium chloride, and that's it. Despite this, there are successful brands that sell for four times the price of ordinary table salt based on their claim to be natural, pure, and even health-giving. The pack of one French brand proclaims: "Derived from the Mediterranean Sea, Costa Fine Sea Salt has been obtained by the simple and natural evaporation of water, aided by the warm sun and wind. It is one of the vital components of taste and contains minerals which are essential to our health." Broste salt, another French brand extracted in the same way, miraculously contrives to contain "a rich balance of minerals and trace elements corresponding to the composition of our own tissue salts." Halen Môn "is natural sea salt produced from the fresh Atlantic waters around the Isle of Anglesey." Maldon sea salt omits to tell us that it is sourced from the muddy, shallow North Sea, emphasizing rather "the ancient craft of panning handed down by generations of salt makers." It is "pure," although not so pure that it does not contain a few (unspecified) "valuable sea water trace elements." Even Blythman falls for the hype on this occasion. She feels that sea salt made by evaporation "is produced in a much more natural manner" than table salt, even though the latter is frequently extracted from underground as brine and then subjected to the very same process of evaporation.[8]

Salt from other sources cannot compete with the romance of the sea, although one brand aims to meet the challenge with the claim that its product, mined in exactly the same way as ordinary table salt, is "millions of years old." What benefit is implied by this claim to antiquity is unclear. Italkali's iodized Sicilian salt tries to have it both ways. "The natural purity and richness of Sale di Sicilia with the added benefit of iodine," it announces, a "healthy and natural" product pure as the day it was laid down in "subterranean Sicilian salt deposits." So where does the iodine come from? Potassium iodate, a chemical related to weedkiller, is added in the factory.[9]

How is it that salt on its own can be claimed as healthy while a small percentage of it in a meal is cause for alarm? There may be traces of valuable minerals naturally present, but this does not alter the fact that it is essentially pure sodium chloride. And as Elisa Zied of the American Dietetic Association points out, even products that claim to be "low sodium" or "unsalted" often contain more salt that we need.[10]

There are alternatives to salt that do contain less sodium. Some cite medical studies adducing their benefits on the pack. But others are committed to the familiar copywriters' nonsense. A product called Lo-Salt contains one-third of the sodium of ordinary salt because it substitutes the similar-tasting potassium chloride. Or, as the manufacturer cunningly puts it, "natural potassium" in exchange for "sodium salt." The implication that one chemical element is somehow more natural than another takes this deceitfulness as far as it can go.

A Dead Duck

"Bird flu won't wait"
—*The New York Times*

It was in 2004 with a flock of 7,000 chickens found in south central Texas that the U.S. faced the crude reality of Highly Pathogenic Avian Influenza (HPAI), better known as the bird flu. Luckily enough, the strain for which the flock of chickens tested positive, (H5N2) was not the same strain (H5N1) that caused the deadly outbreak that has stricken Asia, Europe, and Africa and has led to dozens of human fatalities.

The story of bird flu as a real hazard to human health began in 1997 in Hong Kong, where 18 people were infected and 6 died after coming into close contact with birds at the city's live markets. The outbreak took epidemiologists—the scientists who study the spread of diseases—by surprise as it had been thought that the virus could not jump from birds to humans without the help of an intermediary species such as pigs. Despite the unpreparedness, further spread of the disease and loss of life was prevented by the rapid cull of all poultry in the city. In 2003, the H5N1 virus resurfaced, now slightly altered and more dangerous, "like the doomsday bug in Michael Crichton's old thriller, *The Andromeda Strain*," according to Mike Davis's warning book, *The Monster at Our Door*.[1] This time several members of a Chinese family died after having visited relatives who kept chickens. Similar cases began to be reported from Vietnam, Thailand, and Indonesia, and the World Health Organization announced a pandemic alert.

Through the autumn and winter of 2005, human cases spread westward, recorded with mounting agitation by the European press. Russia, Kazakhstan, Kurdistan, and Turkey were hit in turn, as the European Union banned the import of live birds and feathers. All the human casualties were found to have been in close proximity to birds—five Azerbaijani teenagers died after plucking a dead swan. Shortly before the Cellardyke swan made its final landfall, a German cat was found to have contracted the disease; cat-owners were told to keep their pets indoors. In the States, fear of the virus led to the creation of the Avian Influenza Coordinated Agricultural Project (AICAP), a University of Maryland–based project to research and help prevent and control avian influenza. It received a USDA grant in 2005 and a $5 million renewal in 2008.

Why such expansive coverage for a disease that had killed a little over a hundred people around the world in three years—less than a quarter of the number who die annually in the U.S. from resurgent tuberculosis? The main reason may be what the media habitually refers to as the "specter" of the influenza pandemic that struck at the end of the First World War, perhaps infecting as many as 1 billion people and claiming 60 million lives. There have been human flu pandemics since—in 1933, 1950, 1957, 1968, and 1977—but none as virulent as in 1918. Add to this the vague, residual fear that what we sloppily call "flu" may not be simply the common cold that happens to be heavy enough to allow us to claim a few days off work but something altogether more serious.

Of course, the media thrives on the unrealized threat of mortal disease that could wipe out entire, well, readerships. Bird flu is just the latest of these—after SARS, MRSA, and HIV/AIDS. But in the case of H5N1 flu, august public health officials have added fuel to the flames, pronouncing that a killer pandemic is "inevitable." Epidemiologists merely use the word "inevitable" to mean that "there were several serious flu epidemic/pandemics in the twentieth century, the virus is able to evolve into new forms against which

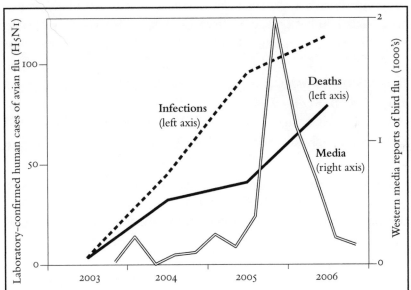

Global human infections and deaths due to H5N1 avian flu and Western media coverage of the virus. Note how media interest is higher during winter months and tails off despite increasing global mortality.
Source: WHO/Factiva database (UK, US, and European news and business publications searched for bird flu).[2]

we have no effective vaccine or natural immunity, and so there is no reason to think that the twenty-first century will be different." Or, as the *New York Times* shrieked, "What is the next plague?" Now it is possible—just about—to make out that the *twentieth* century was a "century of viral pandemics," although few would define it in those terms despite the devastating advent of HIV. There is no reason to think that earlier centuries fared better, and so none to think that the twenty-first century should not have its share of viral outbreaks too. The question is whether outbreaks will be more frequent or more severe to earn the century this label.

This kind of language is symptomatic of an old tendency to blame ourselves for visitations of pestilence. But here it has a modern dressing of anti-capitalism. For Mike Davis, SARS and HIV were the "deadly by-product of a largely illegal international wildlife trade, intimately connected with logging and deforestation, which mortally threatens human health as well as regional biodiversity," while avian flu has prospered "in ecological niches recently created by global agrocapitalism."[3] Michael Gregor, MD echoes him in his book, *Bird Flu: A Virus of Our Own Hatching.* So it's all our own fault, then. It's our affluent lifestyles, our traveling, our migration to the cities that's doing it.

But it is clear that modern life is not mainly to blame. It is quasi-rural backyards with small flocks of chickens where people catch bird flu, not intensive poultry sheds. On the whole we enjoy better food and personal hygiene and experience less risk of picking up an infection from a non-human source than in the past. Even where the risk is still high, the solution through measures such as clean water supply is within our grasp if we are prepared to pay for it.

It is true that an outbreak today is more likely to be global. But sheer geographic spread does not mean that the virus responsible is uncontainable. The pneumonia-like Severe Acute Respiratory Syndrome (SARS) emerged in November 2002. Over 8 months, it spread to 26 countries. In all, more than 8,000 people were infected, of whom 774 died. In the most dramatic illustration of the way disease can spread in the global village, a doctor who had attended to some of the first victims in China traveled to Hong Kong for a wedding. There he passed the infection to 16 other guests. Within a day, those people and others whom they infected in turn distributed the virus to half a dozen countries as far afield as Ireland and Canada, ultimately accounting for 355 cases.

The SARS episode now provides an exemplary case study of how a powerful virus spreads—and may be contained. For although SARS did rapidly kill 3 times as many people as avian flu had done, it was contained relatively efficiently by isolation and quarantine of those infected.

Preparedness in this case did not mean laying in vast stockpiles of a vaccine since SARS was a completely new virus. Nor would some of the usual government reflex gestures, such as closing borders, have made any difference in time. The key was openness and rapid reporting, rather than the sort of administrative denial that can give a virus time to spread unchecked. Another lesson applicable to pandemics of all kinds is the importance of accurate diagnosis and decisively taken preventive measures. In general, even very contagious diseases may be controlled by means of appropriate education as much as by high-tech medicine.

If a flu pandemic is indeed "inevitable," it is equally inevitable that it will start in Asia. It is the high "viral load" here that tempts epidemiologists to use the word in the first place. All of the modern global flu outbreaks began in China, where many people live crammed into unsanitary conditions, frequently cohabiting with their ducks, chickens, and other livestock. The future issue for the rest of the world may be containment; the issue in rural Asia now is domestic hygiene.

Despite this reality, western governments and media continue to play to the gallery. The viral load is nowhere lower than in the United States, yet this is where some of the most hysterical coverage is seen. The *New York Times* alarmed its readers with news that the vaccine being stockpiled by the U.S. "protects only about half the people who receive it," making it sound for a moment as if survival would be a lottery even for the inoculated. Only later in the story came the admission: "The disease has not reached the Americas," but even after that fact was when the paper doggedly refered to the virus as "a pandemic that wasn't but might be."

In Britain, the *Sunday Herald* leaked a government study predicting the breakdown of society: "A minimum of 25 percent of the population will become ill over each six- to eight-week period . . . Mortality is likely to be high—estimated at 1 percent of the total population."[4] Not that there was much to be done about it; nothing yet needed doing. For anywhere else, the precautions were elementary. The UK National Health Service advised people traveling to affected regions to avoid close contact with poultry and to wash their hands. Washing your hands is never a bad thing, of course, but the clueless advice hardly seems commensurate with the claimed scale of the looming disaster and media talk of stockpiling body-bags. It recalls that of the hygiene-obsessed economist Edwin Chadwick, who went to his grave restating his belief in "soap and water as a preventative of epidemics" long after the celebrated John Snow had narrowed down the source of the 1848 London cholera outbreak to a particular street pump.[5]

So far, so terrifying—if you live in the UK. Perhaps it is more instructive to look at what you have to do to catch this disease rather than what to do to prevent it. What does it take to catch the flu from a bird? Scientists discovered in 1974 that the avian flu virus thrives in the guts of wild birds, notably waterfowl. Unlike familiar forms of human-transmissible flu, which is a respiratory infection passed through the air, avian flu is passed on when healthy birds ingest water containing the excreta of infected birds. For a person to become infected, he or she must have intimate exposure to the excreta or intestines of an infected bird. Taking up voodoo and smearing yourself with the raw entrails of an infected chicken in a sacrificial ritual would thus be a good way to contract bird flu. Another successful method is to eat (carelessly prepared) raw ducks' blood sausage, which is a Vietnamese delicacy and the cause of some cases in that country. Sharing your accommodation with poultry is hazardous because the birds' excreta or remains are likely to contaminate human food. This explains the majority of human cases of bird flu—no cases have been recorded of people catching the disease from live wild birds or from cooked poultry.

So even in the UK, it is statistically true to say that you are more likely to die of rabies than bird flu simply because one man—a professional bat handler—died of bat rabies in 2002. This comparison may seem facetious, but both avian flu and rabies are—along with HIV, Ebola, and measles—zoönotic diseases, that is to say they are carried by animals but can pass to humans if the virus mutates in the right way. The comparison illustrates the fact that for both these normally animal-borne viruses, human beings have to take extraordinary steps to put themselves at such a disadvantage that infection is possible.

Even then, the avian flu virus must be present in a form that is able to attack human cells. There are 144 possible combinations of the sixteen H (haemagglutinin) and nine N (neuraminidase) chemical groups that dot the surface of a flu virus. H5N1 is one particular combination. The previous flu epidemics of the twentieth century were H_1N_1 and H_2N_2 and H_3N_2. Other combinations arise from time to time—for example, in the Netherlands in 2003 an outbreak of H_7N_7 led to a cull of 11 million chickens. One person died out of 83 infected by this less deadly strain. But scientists do not know what mutations are needed to produce a dangerous type of the virus. They cannot predict when mutations will appear or what they will be. This makes it impossible to prepare specific vaccines or anti-viral treatments before the new type is present in humans.

It is this mutability that makes the flu dangerous. The smallpox virus, for example, has a fixed composition, which has made it relatively easy to eradicate. The H and N groups of the flu virus, on the other hand, may "drift" into a different form if the amino acids within them are altered. Darwinian natural selection then ensures that the virus reappears each year in a slightly different version, although this change happens so slowly that the human immune system is generally able to cope with the new invader. But the virus can also "shift" rapidly into a new type when a new H or N is introduced, for example, from birds. When this happens, there is no ready immunity in the general population.

"Mutation" is also a word made for scare stories. It seems to leap straight from the pages of John Wyndham, where some malevolent biological slime maneuvers for world domination using processes clearly outside human understanding or control. Mutation is an evolutionary process, but it is not directed. The flu virus is not out to get us; it is simply out to survive, which it already does in the bird population. A sequence of mutations is required for H5N1 to be able to thrive not in a bird's gut but in the quite different conditions of the human respiratory tract. These mutations may or may not occur. And if a human-adapted virus does result, it may or may not turn out to be highly pathogenic.

Paradoxically, a virus that is both highly infective (i.e. it spreads efficiently) and highly pathogenic (i.e. it kills a high proportion of those whom it infects) may not be very dangerous on a global level because it quickly kills too many of its host species to be able to continue spreading. The pandemic threat comes from a virus that is only moderately pathogenic but highly infective. The 1918 flu was such a virus, infecting more than half the world's population but killing only one in twenty of those it infected. The H5N1 virus circulating today is highly pathogenic (it has killed more than half the humans it has infected) but much harder to catch. If it mutates into a highly infective form, that in turn is likely to make it ultimately *less* lethal worldwide.

Recorded outbreaks have claimed single victims or small groups of people, almost all of whom can be directly linked to birds. All of this means that it is extraordinarily unlikely that, for example, Princess Diana's driver was suffering from bird flu on the night of her fatal car crash, a theory reportedly entertained by the *Daily Express*. Bird flu remains essentially a disease of birds even though it may have taken the *New York Times* eight paragraphs in its story about the virus "spreading rapidly through Asia, Europe, and Africa" to remind its readers of this basic fact.

There are other reasons to worry less about bird flu. The media has repeatedly said that a pandemic is "overdue" or even "long overdue." This claim is not based on any virological dictum, but simply on the historical pattern of outbreaks—one every ten to fifteen years or so from 1918 to 1977 and then a pregnant pause until now. But in fact the longer H5N1 "tries," the *less* likely it is to succeed in adapting into a human-transmissible type. Some scientists now believe H5N1 has had its chance. A virus is not like a volcano, where pressure may build up gradually, leading to an eventual eruption. And a continued non-outbreak doesn't make a future outbreak more likely or more deadly.

It seems these truths are at last being acknowledged. In February 2007, H5N1 avian flu rampaged through a Bernard Matthews turkey shed in Suffolk, requiring the swift destruction of 160,000 birds. Media talk of an inevitable human pandemic promptly vanished, replaced by a new spirit of optimism. When push came to shove, the outbreak was simply something that had to be—and could be—dealt with.

While bird flu has yet to claim a single human victim in Europe or the Americas, and has killed fewer than 300 people worldwide, it is perhaps worth adding that the familiar winter flu that nobody panics about claims at least 30,000 American and 12,000 British lives each year.

It's Amazing What They Can Do

"Concern mounts as bacteria resistant to antibiotics disperse widely."
—*The New York Times*

Where would you go to catch a dangerous infection? A crowded commuter train, an airplane, an infant's school, a brothel? How about a hospital?

"MRSA, a drug-resistant bug, lurks in Washington hospitals," warned the *Seattle Times* in 2008, exposing the shocking rise in numbers of victims, from 141 a year to 4,723—numbers the paper claimed had been suspiciously witheld from public documents by said hospitals. Other media sources labeled MRSA a "super bug" and even gave it an odd "bug of the year" award.

A couple of years prior, Britain was seeing similar reports: "Hospital bug that kills in 24 hours," warned the *Daily Mail* in December 2006 as an apparently new strain of the already familiar "superbug" MRSA claimed its latest victims: a nurse and a patient in Staffordshire. "Hospital crisis as PVL bug kills tot," added the *Sun* five days later as a premature baby boy died in Norwich.

PVL (Panton-Valentine leukocidin) is a toxin sometimes produced in association with MRSA. It destroys the white blood cells that the body needs to fight infection and can quickly kill by causing pneumonia and other conditions.

MRSA (methicillin-resistant *Staphylococcus aureus*) was identified in the 1960s as a variant of the *Staphylococcus aureus* bacterium that had developed resistance to methicillin, an antibiotic introduced after the bacterium had already begun to show resistance to penicillin. Occasional MRSA infections were observed from this time, but newspapers began to use the word "superbug" to describe MRSA in the 1980s, when outbreaks began to increase, especially in American and British hospitals. In 1995, in New York City alone, 7,800 people became infected with drug-resistant *Staphylococcus*; 1,400 of them died.[1] *Staphylococcus aureus* mainly infects open wounds, causing boils and abscesses, and is therefore especially dangerous to surgical patients. It is this that makes it a hospital disease.

So are hospitals now places to be feared rather than places of healing? An extensive report by the *Journal of the American Medical Association* revealed that there had been 8,987 observed cases of the bacterial infection in the U.S. in 2005—doubling the previously reported numbers. Of the MRSA-related deaths in that year, 1,598 occurred in hospitals. According to the report, the virus was killing more people each year than HIV/AIDS. The British media have widely quoted a figure of 5,000 deaths each year due to MRSA as estimated by the National Audit Office. But another government organization, the Health Protection Agency (HPA), calls this figure "incorrect," pointing out that it describes deaths from *all* hospital-acquired infections. Many people thought that MRSA was the major, if not only, such infection, hence the confusion. It is cold comfort that the death toll stands and that there are simply other infections to worry about.

However, deaths are not a very reliable indicator of the prevalence of MRSA. The numbers may be an underestimate if death is still ascribed to a familiar killer such as pneumonia rather than MRSA where both are present. On the other hand, better diagnosis undoubtedly accounts for some of the rise in recorded MRSA, while MRSA is much more likely to be mentioned on death certificates of people who died in the hospital than on those of people who died elsewhere could arise from an excessive readiness to name MRSA because it is already associated with hospitals.

Meanwhile, the newspapers have grown shriller as nature has come up with new tricks. Unlike ordinary MRSA, which affects the most vulnerable, PVL has the impressive property that it afflicts the "young and healthy"—among them, therefore perhaps even more of their readers. "Remember these initials," intoned the *Independent*, putting them in letters three inches high on its front page just to make sure that readers did. The first PVL fatality during this period was an 18-year-old who died after infection of a leg injury sustained while he was on a commando training exercise.

The story of schools as breeding grounds for the infection particularly interested the U.S. media. Following the death of a Brooklyn middle school student in April 2005, the *New York Times* reported that "Across the country, classrooms were closed for disinfection as politicians and health officials scrambled to calm an anxious public. Suppliers of sanitizers and disposable blood pressure cuffs seized the moment to gin up demand." Keen to find a sensational angle, Fox News erroneously reported that the gay community, especially San Francisco's, was at heightened risk. The news network later offered a formal apology.

Microbiologists had long warned that consequences were to be expected from doctors' enthusiasm for prescribing antibiotics, as the bacteria they were designed to combat gradually developed resistance. At first, the fear was of one "doomsday superbug," probably MRSA.[2] But more recently, it has seemed that we will come under attack by a whole range of bugs. *Enterococci*, *Clostridium difficile*, and tuberculosis are among the new or resurgent bacteria that have acquired resistance to antibiotics.

In January 2007, the *Guardian* issued a "Superbug 'apocalypse' warning." The article that followed quoted Professor Richard James, the director of a new Centre for Healthcare Associated Infections at Nottingham University: "We are facing a future where there will be no antibiotics and hospitals will be the last place to be if you want to avoid picking up a dangerous bacterial infection—in effect, cut your finger on Monday and you'll be dead by Friday if there's nothing to prevent it."

This pessimistic vision carries with it a sense of the medical profession's profound dismay that the antibiotic revolution trumpeted as a miracle scarcely half a century ago now seems to lie in ruins. Yet some of the reasons for this are to be found with the medics themselves. Like the rest of us, many doctors believed that antibiotics were indeed a miracle cure. The use of antibiotics in major surgery as well as in the treatment of infections of many kinds has unquestionably contributed to the public's grateful awe of doctors found in the popular expression "It's amazing what they can do."

Although it had been anticipated—and later was shown to be true—that bacteria would respond to hostile antibiotics by adaptation to produce more vigorous strains, it was easy to hope that antibiotics would remain good for all time like ordinary chemical medicines. Frequent indiscriminate and precautionary prescribing of antibiotics merely hastened this process of evolution. In hospitals, meanwhile, according to Laurie Garrett's book, *Betrayal of Trust*, "'Typhoid Mary' doctors" have also been directly responsible for spreading infection from patient to patient because they are too hurried or too arrogant to wash or submit to testing.[3]

While many doctors have held unrealistic expectations of antibiotics, we have all, perhaps, cherished unrealistic expectations of what hospitals can do. The idea of public health was born when hospitals were "little more than warehouses for the dying," according to Garrett.[4] During the third quarter of the twentieth century, they became rather more than that, but this period currently appears to us more like a brief interlude than the latest stage of an upward progress. In terms of infectious disease, hospitals may be reverting to type.

The reasons advanced for this to reflect the range of political prejudices about how health services should be run. The transfer of power from the public health authorities to the medical profession in the United States and the reorganization of public hospitals as trusts in Britain are among the suggested factors. The introduction of privatized services, including cleaning, coincides unfortunately with the rise of resistant bacteria. Hospital mergers and centralization may also play a role, since larger hospitals tend to have disproportionate levels of MRSA, while closures facilitate the spread of superbugs as workers relocate.

A full hospital that turns beds over rapidly presumably restores more people to health than a half-empty one. But high occupancy rates are also thought to help the spread of MRSA. "Headlines about patients dying while waiting to go to a hospital have given way to reports of patients dying simply by staying in one," as one newspaper tartly put it.

MRSA levels are high in the United States and Britain, comparable with France but twice the level in Germany or Spain. The situation in all these countries has been compared unfavorably with that in the Netherlands and Denmark, where MRSA infection is extremely low. However, some countries have little MRSA but more of other hospital-associated infections, and it is not clear that anybody has a complete answer. Dutch hospitals operate a praised "search and destroy" policy whereby patients are screened for MRSA on admittance and placed in isolation wards if necessary.

It is tempting to seek parallels for the MRSA phenomenon in previous scares involving bacteria, such as salmonella and listeria in food, as well as diseases such as BSE in beef cattle or the threat of avian flu, and a generalized fear of invisible germs. But this is to neglect the hospital connection that is so much a feature of the media coverage. Consider instead that the fear of hospital-associated infection may mask a deeper fear—that of having to go to the hospital at all. The dread of surgery in particular, though seldom discussed, ranks among the highest public fears, according to Paul Slovic's *The Perception of Risk*.[5]

Another distinctive factor in the MRSA story is that at least part of the solution to the crisis is widely thought to lie with improved hospital hygiene. Unlike the transmission and treatment of infectious diseases, this is a topic on which everybody has an opinion. We see with our own eyes that many hospitals are filthy. The issue of cleanliness, which emerged as MRSA infections climbed in the late 1990s, gave the media and the public a campaigning focus of both practical and metaphorical value. Hospitals needed to be cleaner, but the way they were run could use a "clean-up" as well.

The obsessive focus on hygiene by the media, and subsequently by politicians with an eye for cheap measures that might make a visible difference, has disguised the fact that the other part of the solution to bacterial resistance must be the creation of more effective antibiotics. For all the use of words such as "doomsday" and "apocalypse," there is no destined biological end point where all bacteria become resistant to all antibiotics. Instead, we need to develop new antibiotics to keep pace with the emergence of new resistant strains and at the same time remain disciplined in prescribing them only in specific circumstances so as not to spur the emergence of resistance. Sadly, this means letting go of the notion that the antibiotics of the last century are everlasting miracle cures and adjusting to the fact that bacteria adapt and change.

It's amazing what they can do. The bugs, that is.

Completing the Course

"Vaccine-autism question divides parents, scientists"
—*CNN*

The triple vaccine for measles, mumps, and rubella (MMR) was the basis for one of the biggest recent scare stories in the United States. The story blew up not when the vaccine was introduced in 1989, but ten years later, when a British doctor published a paper that appeared to link the vaccine to autism and bowel disease.

The MMR vaccine was designed to protect against three viral diseases of which older readers may have direct memories. Measles is a highly contagious disease characterized by an itchy rash. In developed countries, the mortality rate is 1 in 3,000, but it can be as much as 1 in 10 where malnutrition is rife. Measles deaths have fallen sharply this decade but, according to figures from the World Health Organization, still claimed 242,000 lives in 2006 among unvaccinated populations worldwide. Mumps produces glandular swelling and is generally less serious, but both mumps and measles can produce life-threatening complications. The principal risk from rubella is to women in early pregnancy whose babies may be born with birth defects.

These diseases have been virtually eradicated in the West with the introduction of vaccines in the 1960s and 1970s. However, a large proportion of the population must be vaccinated in order for society as a whole to be protected. For the measles vaccine, coverage required for the population, as a whole, to be safe from an epidemic (to reach what is called herd immunity) is 90 percent or more because the disease is so contagious. Measles remains a major killer in developing countries simply because the vaccine doesn't reach enough people. Because it is hard to achieve the necessary coverage, and because not all vaccine doses are effective, immunization programs frequently recommend a second dose, which greatly reduces the chances that a given person does not have at least some protection.

The measles-only vaccine introduced into the U.S. in 1963 is purported to inoculate 95–98 percent of recipients in the first dose, and 99 percent in the second dose.[1] In the first 26 weeks of 1989, just before the mass proliferation of the MMR vaccine, there were 7,335 notified measles cases and at least ten deaths (with an additional 30 deaths strongly suspected of being measles-induced) in the U.S.[2] During the first 20 years of the vaccine's use, it prevented an estimated 52 million cases of measles and 5,000 deaths.

What happened next was the derailing of the juggernaut of medical progress toward the eradication of measles in the United States and beyond. In February 1998, Dr. Andrew Wakefield, a gastro-enterologist at the London's Royal Free Hospital, published results of tests on 12 children with both intestinal inflammation and developmental disorders, including autism. The paper mentioned work by others (and previous work by Wakefield on the measles-only vaccine) in which the viruses and/or their vaccines were "associated with" or "implicated as risk factors" for autism; it left similar doubt hanging in the air over the MMR vaccine. "We did not prove an association between measles, mumps, and rubella vaccine and the syndrome described," the authors wrote in the *Lancet*.[4]

Not only did Wakefield not prove an association, he did not produce *any* evidence to support one. Wakefield simply wrote: "Onset of behavioural symptoms was associated, *by the parents*, with measles, mumps, and rubella vaccination" (italics added). That was all. There the story surely would have died except for the fact that at a press conference, Wakefield pursued this idea of a link and suggested that parents might wish to have their children put through three courses of separate vaccinations, rather than undergo the triple vaccine.

The story hit the headlines immediately, but grew only slowly in ferocity for reasons largely unrelated to the quality of the original science. "Vaccine Jitters" was the title of *Time* magazine's September 1999 cover story in response to *Lancet*'s report. Simultaneously, new work appeared to refute Wakefield, and the *New York Times* countered with: "Doctors fighting backlash over medicines."

Starting in 1998, worried parents not only sought single vaccinations against the three diseases, or vaccination solely against measles, but also shunned other vaccines that had not yet been implicated. In the UK, the *Daily Mail* took up aggrieved parents' cause, championing their attempt to win the right to choice in court: "Parents shun MMR jab over health fears"; "Jab damaged our children, say 2,000"; "Parents' fear as the single measles jab is withdrawn. Drug company denies giving in to government pressure."

The scare was unusually effective for a number of reasons: The public trusts what doctors say, but does not like being told what to do by governments; parents seek their children's safety and well-being; people naturally fear immunization; and autism was apparently on the increase. The numbers of people identified with autism had begun to rise in the 1970s in the United States and in the 1980s in Britain, largely coincident with the respective dates of introduction of the MMR vaccine. Autism was doubling every five years, partly because of better diagnosis and more inclusive definitions of the condition, but partly, it seemed, because of unidentified, possibly environmental, factors. Identifying links with other conditions might help scientists understand the causes of autism, and it was understandable that people should light upon one potential link where it was easy to lay the blame.

The immediate consequence of the press campaign was a fall in immunization levels, especially in the UK, where immunization is mainly optional. No evidence has been presented to support Wakefield's theory, while many large studies, including one in the *Lancet* as early as 1999, could find no link. These culminated in a review of 31 studies, each involving hundreds, and in one case half a million, subjects, by an international team for the Cochrane Library, an organization set up to evaluate medical research.[5] This received only a fraction of the coverage of Wakefield's study of 12 children.

As evidence grew refuting Wakefield's claim, Wakefield himself was accused of having been paid £55,000 (almost $80,000) by lawyers for the families who had brought their children to him, an interest he had not declared to the *Lancet*. Most of his co-authors publicly dissociated themselves from him.

Parents with children of immunizable age could be excused for being muddled about all this. In any case their stories seldom fit into neat clinical slots. So in order to imagine their predicament, it may help to consider some examples. (The stories are true, although some details have been changed.)

It's 1999. William, a bookie, has his first addition to the family. He remembers having measles as a child and, though it didn't kill him, he is a little worried by what he has read about the MMR vaccine. He notes that MMR coverage is holding up and reckons his child will be safest of all unimmunized but reliant on herd immunity.

It's 2001. Fiona, a florist, has a two-year-old son beginning to show signs of a developmental disorder. The boy had his first MMR shot a year ago, but now she must consider whether his young sister should have her first shot in a few months' time. The boy's disorder may be genetic, but then again, perhaps the MMR was responsible. She decides the girl should have the vaccines singly in a private clinic. As the doctor gives the shot, he vouchsafes that his children had the combined vaccine, and he thinks it's safe.

It's 2002. Juliet, a busy actress, has a daughter who had the MMR shot and now must decide whether to do the same with her 18-month-old son. She remembers reading an article a couple of years ago in a British satirical magazine recounting the conviction of a parent whose child had developed autism that it was the MMR vaccine that must have caused it. Her "blind faith" in medicine is rocked. "I had learned how vigilant you have to be as a parent because you are responsible for this little life." She books her boy in to a private clinic to have the three vaccines separately: first measles, then rubella, and finally mumps.

How did these children fare? William realizes that protection against contagion is actually frighteningly low where he lives—against Department of Health advice, people are holding "measles parties" so that their unprotected children might get the disease over with. He belatedly puts his child in for the MMR shot. Fortunately, the child has not caught measles in the interim. Fiona's daughter is protected against measles and did not develop her sibling's disorder. Juliet (who is in fact the actress Juliet Stevenson) duly takes her son to a crowded, disorganized clinic for the measles and rubella shots. By the time that she is free of work commitments and can think about the final shot for mumps, she calls the clinic only to find that it has closed some months ago. The doctor who ran it has been arrested and charged with forging the results of blood tests. An independent test shows that Juliet's son actually has no protection against measles. "Now I don't know what to do. Do I get him single measles again, or forget the whole thing?" she wonders aloud to the *Mail on Sunday* while promoting her role in a cathartic drama about the dangers of MMR.

As the example of mothers like Juliet Stevenson shows, the relative safety of a vaccine shown by statistical evidence is not always persuasive. Vaccination allows a pollutant to enter the body and so breaks a deep taboo. Vaccines do not work in all cases and produce adverse reactions in some. For these reasons, people have feared vaccination since its invention—an occasion commemorated in a lurid caricature published in 1802 by James Gillray showing people receiving Edward Jenner's smallpox vaccine erupting in bovine buboes.

This timeless fear is overlaid by characteristically modern concerns that seem bound to surface afresh with each new vaccine introduction. "Vaccines come under greatest scrutiny when they are successful," as one doctor pointed out at the height of the single vaccine rush. When they are the only alternative to a nasty disease, they are uncritically accepted. But today we—including many of our doctors—have forgotten what it is like to suffer, or witness suffering, from many of the diseases that we now immunize against. Smallpox was the biggest killer disease in the West during the eighteenth century, tuberculosis in the nineteenth, both eradicated by vaccination.

Without unpleasant reminders of what vaccines are for, we have the luxury of speculating about side effects and links to other disorders. Combined vaccines—which are likely to become more of a feature of immunization in the future—mean fewer shots, but excite fears of chemical "cocktails" and overloading children's immune systems. As Stevenson told her interviewer, "There are hundreds, if not thousands, of parents in this country who deeply believe that their children were damaged by this vaccine, yet they have still not been given credibility." But most parents are not immunologists, and while the depth of their belief is not in question, its basis in reliable observation most definitely is.

Autism is an obvious focus for parents' fears because it is often first observed at an age shortly after children have had their first immunizations. (Before MMR there were fears that other vaccines might be linked to autism.) However, a closer look at the timing of autism diagnosis in relation to immunization reveals an interesting pattern, with parents tending to spot signs much sooner (around six weeks) than health professionals (six months or more).[6] Yet if there were a link between MMR and autism, one would expect the professionals to notice the signs as soon as the parents. Alternatively, if there is no link, one would again expect everybody to notice signs at the same time—when they appear. The fact that the experts observe the signs later could conceivably point to professional conservatism, incompetence, or denial. Perhaps the parents' earlier observation can be attributed entirely to natural anxiety. Sadly, this does not mean that the autism is not real, but it does suggest that the link with the MMR vaccine is imagined.

Blaming the MMR vaccine for autism rather than one's genes, or sheer bad luck, is a complicated displacement that implicates science and technology, the medical establishment and modernity in general. But it is also a pernicious way of blaming ourselves—we as parents put the child to the needle—and it is this tendency above all that the media exploits so cynically in scare stories of many kinds.

The effect of the anti-MMR campaign was to add to the defaulters who did not have their children immunized out of ignorance or idleness a large number of "active resisters," who were, according to Michael Fitzpatrick, a doctor who wrote a parents' guide to MMR and autism, "middle-class, well-educated parents who had chosen not to have their children immunized."[7] These parents pitched themselves against a government that refused to give the option of single vaccines, as is sanctioned in some other countries. The combined effect of these two groups pushed immunization well below the safety threshold for herd immunity.

It began to seem that matters would only be resolved the hard way. Cases of measles, mumps, and rubella in the U.S. all increased slightly between 2003 and 2005.[8] In the UK, where the scare was more widely perceived, the number of measles cases reached nearly 1,000 in 2007—a level not seen since before introduction of the triple vaccine, and more than double the number of cases seen in the U.S. in the same year. In March 2006, a thirteen-year-old boy became Britain's first child measles fatality since 1992.

The year 2004 did bring some respite to anxious parents when news sources began widely—and confidently—reporting findings that ran contrary to Wakefield's original claim. "Researchers retract a study linking autism to vaccination," the *New York Times* reported in March 2004, referring to the public admission by ten of Wakefield's co-researchers that their findings had been insufficient to prove a link between the vaccine and autism. At last, the debate came to an apparent close in early 2009 when an American special court ruled against parents seeking compensation for their children's alleged vaccine-induced autism. The court dismissed the case on the grounds that there was no scientific correlation evident.

Thankfully, the MRR vaccine is in the clear for now, but there are always plenty of others the scientists and media can worry us over—as they almost certainly will.

Sudden Death

"Researchers link mutant gene to crib deaths in blacks"
—*The Los Angeles Times*

Myriad worries face parents of newborn children—not least among them the fear of an unexplicable death. Sudden infant death syndrome (SIDS) is any sudden and unexplained death of an apparently healthy infant aged over one month. The term "crib death" is often used in North America and "cot death" in the

UK. The diagnosis of crib death is a so-called definition of exclusion, applying only to an infant whose death remains unexplained after the performance of post-mortem investigations, including an autopsy. Generally, the infant is found dead after having been put to sleep and exhibits no signs of having suffered. The fear of sudden infant death means that new parents regularly visit their sleeping baby to check that he or she is still breathing.

The extent of parental worry does not make much sense as SIDS is responsible for very few deaths. In the U.S., it accounts for roughly 50 deaths per 100 thousand births. The rate is a little higher, around 60, in Scotland, and a little lower, around 40, in England and Wales. This is far fewer deaths than caused by congenital disorders and disorders related to short gestation, though it is the leading cause of death in otherwise healthy babies aged one month to one year, albeit a category in which there are very few deaths. Moreover, the incidence of sudden infant death syndrome has fallen in all developed countries in recent years, in some cases dramatically. A fall of nearly 90 percent in Australia over 15 years is particularly noteworthy.

But the worry remains very real, reflecting both the horror of losing a young child and the "incomplete" medical explanation for the event when it does occur. Very little is known for sure about the cause of SIDS, and there is no proven method for complete prevention. That said, many studies of babies around the world that have died suddenly and unexpectedly have identified a number of risk factors. Prenatal risks include inadequate nutrition, a smoking mother, teenage pregnancy, alcohol and hard drug–abuse, and an interval of less than one year between subsequent births. It is striking that the rate of unexplained infant death in England and Wales is 4 times greater for children born to mothers aged under 20 than those born to mothers in their thirties, and 6 times higher for those children registered at birth only to the mother than for those children born to a married couple. Studies in the U.S. have also shown that African American and Native American infants are 2 to 3 times more likely to die from SIDS than white infants.

Postnatal risks include low birth weight, exposure to tobacco smoke, excess clothing and overheating, excess bedding, soft sleep surface and stuffed animals, and sharing the parental bed. It also seems that boys are more likely to suffer than girls—nearly two-thirds of deaths are accounted for by boys. Putting young children to nap in their own bedroom is also thought to be a risk.

Laying an infant to sleep on his or her stomach is widely thought to be one of the greatest risk factors. Accordingly, many governments have introduced "back to sleep" campaigns, encouraging parents to put their children to sleep on their backs. There are various theories supporting the risk of prone sleeping. The first is the idea that small infants with little control over their heads may, while face-down, inhale their exhaled breath or smother themselves on their bedding. A second theory suggests that babies sleep more soundly placed on their stomachs and are unable to rouse themselves when they have an incidence of sleep apnea, or breath-holding, which is thought to be common in infants. It has also been suggested that the victims of SIDS might have abnormalities in or delayed development of the arcuate nucleus of the brainstem, contributing to their death. Whenever "back to sleep" campaigns were introduced, in the early 1980s in the UK and the early 1990s in the U.S., SIDS death rates fell, often sharply.

Death might also be caused, some researchers say, by the toxic nerve gases produced through the action of fungus in mattresses on the chemical compounds frequently used to make mattresses fire retardant. A major plank in this explanation is the widely observed phenomenon that the risk of crib death rises from one sibling to the next. After one crib death in a family, the risk of recurrence for a subsequent child is up to 5 times the rate of the population more generally. No satisfactory biological explanation for this has ever been put forward. But the toxic gas explanation fits this neatly as parents generally buy new bedding for their first child, reusing it for later children, with a greater chance that there will be fungus that has become resident in the material, in turn leading to a higher chance of crib death. Single mothers or poorer families might borrow used or buy second-hand cribs, perhaps accounting for the higher death rates among these groups.

A decade ago, the New Zealand government issued advice recommending that new parents either buy bedding free of the toxic compounds or wrap the mattresses in a barrier film to prevent the escape of gases. It is claimed by those supporting this view that no case of crib death has ever been traced back to such a manufactured or wrapped mattress. Of course, face-up sleeping could well reduce the death rate if this is the cause for SIDS, as the dense gases that cause death diffuse away toward the floor—a baby sleeping face-up is less likely to inhale them.

More recently, a brain abnormality has been found in the victims of SIDS that could cause the babies to suffocate if they become smothered by bedclothes. The U.S. researchers claimed that this was "the strongest evidence yet of a common cause for [crib] death." The debate rumbles on.

One particularly sensitive issue is the link between crib death and child abuse. All professionals involved in crib deaths accept that there is a small number of cases where a parent or carer has done something unlawful to contribute to the death. A number of pediatricians have said that they believe some cases diagnosed as SIDS are really deaths resulting from child abuse. Their suspicions are particularly aroused in the case of multiple crib deaths within a family. Indeed a dictum known as "Meadow's Law," after the well-known former pediatrician, says that "one cot death is a tragedy, two cot deaths is suspicious and, until the contrary is proved, three cot deaths is murder."

Sudden infant death syndrome spares no country, but it reached a fever pitch during the 1990s and early 2000s in the UK, when a number of mothers of multiple apparent SIDS victims were convicted of murder. At one of the trials, Sir Roy Meadow, speaking as an expert witness for the prosecution, made one of the most infamous statistical statements ever in a British courtroom. He claimed that the chance of two children in the same affluent, non-smoking family both dying a crib death was one in 73 million. In a complex, confusing, and emotional case, the statement provided something definite to hold on to and was widely headlined in the national press. The mother, Sally Clark, who had had two children die of crib death, was convicted of murder.

Unfortunately, the figure was in all probability wrong and certainly misleading. The statistic was derived from the Confidential Enquiry for Stillbirths and Deaths in Infancy, a study of baby deaths in England in the 1990s. It estimated that the chances of a randomly chosen baby dying a crib death are 1 in 1,300, falling to around 1 in 8,500 if the child is from an affluent, non-smoking family, with the mother aged over 26 years. If—and it is a big *if*—there is no link between crib deaths of siblings, then the chances of 2 children from such a family both suffering a crib death is obtained by multiplying the odds, namely 1 in 8,500 by 1 in 8,500. This produces the probability of 1 in 73 million.

But all the evidence suggests that there are links between such deaths, which are not independent. If the odds of a second crib death in a family are around 1 in 100, the odds of a double crib death would fall to about 1 in 130,000. Since around 650,000 children are born every year in England and Wales, we might expect as many as 5 families on average each year to suffer a second tragic loss. This paints a far less dramatic picture than the 1 in 73 million figure. Interventions from a number of quarters, including the Royal Statistical Society, led to some high-profile acquittals at subsequent retrials. Sally Clark died of alcohol intoxication in 2007. A family statement after her death said "she never fully recovered from the effects of this appalling miscarriage of justice."

The ambiguity, uncertainty, and complexity in defining SIDS incidents has had an impact on the statistics we use to define the issue. Statistically in the U.S., sudden infant death syndrome did not "exist" until it was identified as a disease in 1969, following the Second International Conference on Causes of Sudden Death in Infants held in Seattle, Washington, where the term was coined and the definition established. Prior to that time, we had no idea of the true incidence of such deaths, and that in turn hindered the search for an explanation of those deaths that did occur. (Many of the infant deaths prior to that time were—wrongly—attributed to respiratory disease.) Perhaps due in part to increased awareness and preventative measures, the rate of cases of deaths caused by SIDS dropped 3 times faster during 1990–1994 than during 1983–1989. During the earlier period there were 61,882 SIDS deaths recorded, but between 1990 and 1994 the SIDS rate dropped at an average rate of 5.6 percent per year. Overall, the SIDS rate was 13.9 percent lower during the latter time period.

But defining cause of death is sometimes an imprecise science, and fashions change over time. In 2004, 4,600 infants in the U.S. died for no immediately apparent reason, but less than half of those were indentified as "sudden infant deaths," the majority being labeled instead as "sudden unexplained infant deaths" (or SUIDS). The number of deaths attributed to this cause has been on the rise, reflecting the suspicions that some pediatric pathologists have, which is that parental or adult intervention may have occurred in some of the cases where an infant dies suddenly. The Center for Disease Control and Prevention admits that "Many SUID cases are not investigated, and when they are, cause-of-death data are not collected and reported consistently. Inaccurate classification of cause and manner of death hampers prevention efforts and researchers are unable to adequately monitor national trends, identify risk factors, or evaluate intervention programs."[1]

Several British government reports in recent years, notably the 2004 Kennedy Report, have suggested that the parents of babies whose deaths are labeled as "unascertained" unfairly face stigma that could reflect nothing more than differing practices among coroners. The Kennedy Report recommended that the term "sudden infant death syndrome" should continue to be used where appropriate, with "unexplained pending further investigation" or the broader category of "sudden unexpected death in infancy" being used for all other cases.

Whatever the definitional issues relating to the figures and the resulting trends, the bottom line would seem to be that parents can reduce to next to nothing the chances of an infant dying unexpectedly by putting it to sleep on its back, on a properly wrapped mattress, and with little else in the crib. The odds also seem to favor girls, first born to a thirty-year-old couple compared to the second or third son born to a teenage mother. Nevertheless, while uncertainty about the causes of unexpected infant death remain—and money is thrown at researchers to pursue such a wide range of avenues—the media will stir up concern, leaving parents of young children with a nagging doubt for years to come.

As the children grow up, the parents can leave behind worries of SIDS but can begin to worry about the emerging phenomenon of "adult crib death" or sudden adult death syndrome, which seems to strike mainly those in their teens and twenties. Evidence is emerging that the number of people who collapse and die suddenly without explanation could be much greater than is recorded in the official statistics, with the cause of the tragic deaths remaining a mystery. One in 20 sudden adult deaths (deaths of adults who are apparently healthy) remains unexplained. Campaigners say that this condition needs to be given a "name" so that the problem will be addressed more seriously, as was the case with sudden infant death from the 1970s onward.

3. Passing the Time

Pastimes designed for relaxation can also be panic-inducing. Alcohol is a well-known demon, but art can also kill. Modern technology might threaten the cinema as we know it, but it has made collecting easier.

Art Is Dangerous

"Umbrellas: a year later shock of deaths wearing off"
—*Los Angeles Daily Times*

Artists have always relished the idea that their work is "dangerous." Picasso, Duke Ellington, and Anthony Burgess are among those who have made this claim in as many words. But the artist's idea of what makes a work dangerous is perhaps not quite the same as everyone else's.

Skyscrapers—especially those that are designed with more emphasis on aesthetics than on functionality—are a common fear for many people. Boston's Hancock Tower, widely anticipated as a cultural landmark by one of the nation's foremost architectural firms, became a danger to civilians before the building was even opened. In 1972, window panes began cracking and falling to the ground below, prompting building management to replace the missing panes with plywood and causing papers to jokingly call the skyscraper "the world's tallest plywood building." Thankfully, authorities responded quickly and no injuries or fatalities occurred, but more problems were still to come.

After Hancock's opening in 1976, its upper-floor patrons began complaining about a "nauseating sway" resulting from insufficient stabilization mechanisms. An analysis of the building's engineering revealed that high enough wind speeds could actually topple the building. Damage control was again successful, but the cost of the new reinforcements came in at $5 million. In the end, the problems were resolved and no serious harm was done, but numerous other art projects have not ended so favorably.

When Carsten Höller became only the seventh artist to fill the vast Turbine Hall of London's Tate Modern art gallery in October 2006, his temporary installation was greeted not with the usual gasps of awe but with whoops of glee from most—and trepidation from a few. The *Times* (of London) was most concerned. Was the piece "Art—or accident waiting to happen?" Höller's work, *Test Site*, was essentially a set of five glorified helter skelters, finely constructed in stainless steel and clear plastic. The longest of the slides was 60 yards, and it took just 12 hair-raising seconds to descend through its chutes and spirals. But were gallery visitors taking the ride possibly on "a slippery slope to disaster?"

The story originated on press night, when sliders reportedly emerged from the tubes at high speed with "swollen ankles, friction burns, grazed knees, and bruised elbows." Extra cushioning was hastily added. Nobody was going to worry much about a few injured journalists, but the safety of the public was naturally a concern as the slides were being assembled. Fortunately, the artist was able to offer solid reassurance: "These are built to German safety standards which the British inspectors are very happy with because they have the reputation of being the best in the world."

But perhaps there is good reason to be wary of art. A few months before these incidents, two women had been killed and a three-year-old girl badly injured when the giant inflatable artwork that they were exploring broke free from its tethers and flew off into the air before snagging on a nearby CCTV mast. The work, *Dreamscape* by Maurice Agis, was on display at a park in Chester-le-Street, County Durham. The PVC structure consisted of colorful chambers linked by a network of tunnels. Wearing equally colorful capes, people could wander through the maze, "disappearing" when their location matched their costume. Since

1996, versions of the work had been exhibited in Denmark, Italy, and Spain without mishap, as well as in Liverpool, where it was vandalized with knives.

But the U.S. and Japan both experienced the full gravity of art's destructive potential in 1991. In October of that year, the artist Christo, best known for his colorful wrapping of edifices such as the Pont Neuf in Paris and the Reichstag in Berlin, unveiled a massive installation of specially constructed metal umbrellas on either side of the Pacific Ocean—1,760 yellow umbrellas in California, 60 miles north of Los Angeles, and 1,340 blue ones at Ibaraki in Japan.

The installations in California had been open little more than a week when they had to be closed to the public due to high winds, which were beginning to damage the umbrellas. Crews struggled in the gale with the dangerous job of cranking the giant metal objects into a furled position. People continued to come and see the artwork, however, and a few days later a woman was killed when one of the 440-pound structures blew free of its foundations and crushed her against a boulder, forcing the complete abandonment of the ambitious project. "Christo umbrellas close: crews dismantling project after woman is killed in accident," reported the *San Francisco Chronicle*.

Worse was to come a few days later. As the Japanese umbrellas were being dismantled, a crane operator was electrocuted when the boom of his crane struck a high-voltage power line. "Japan crane operator killed dismantling Christo umbrellas," noted the same paper four days later.

What is the problem with art? Does it really pose a mortal danger? And why is the problem apparently getting worse? The main factors are easy to identify. But the true nature of the phenomenon is obscured by the attitude of the media. Disproportionate coverage of accidents of this type—compared to, say, accidents suffered by members of the public on building sites or while at leisure facilities—reflects newspapers' pretended bafflement over "conceptual" art.

Art in galleries is the province of the critics and may safely be ignored by the rest of the press corps. But when art escapes these boundaries, it becomes fair game for any philistine reporter, especially when the work has been paid for with public money.

The first reason that art goes wrong is that, at some level, it is meant to. Art is risky. Part of its job is to revive in us a feeling of visceral sensory connection with the world, and this in principle includes its dangers. Carsten Höller produced a French sociologist who claimed that one merit of Höller's work was its ability to induce "a kind of voluptuous panic upon an otherwise lucid mind."

After the double tragedy of the umbrella deaths, Christo acknowledged he was saddened by the loss of life that his work had occasioned, but then offered this brazen piece of post-rationalization: "All my works of art are created to challenge normal views of art. [*Umbrellas*] challenged the view that art is safe. They're not make-believe. The risk is real—almost like climbing the Himalayas."[1] It is perhaps curious then that on his personal website, in an otherwise detailed chronicle of the setting up and taking down of the umbrellas, Christo makes no mention of the deaths that he seems to believe validated his work.

Art is different in this respect from architecture. The media tends to regard the more substantial and widespread problem of architectural failures as deserving of serious treatment. With catastrophic building collapses, the press may occasionally rejoice in the misfortune of the architect and developer, but any hazard to the public is taken seriously. This is because, unlike art, where our expectations are uncertain, we expect buildings to perform properly and are shocked if they don't. Where fatalities occur, as in the wave of collapses of sports halls and similar buildings prompted by exceptionally heavy snow in eastern Europe during the winter of 2005–2006, the tone is appropriately somber. The media's handling of building failures only resembles that of its art coverage when the architecture concerned is self-consciously avant-garde and when it is judged that lives are not seriously at risk. Thus, the wobbling of Norman Foster's Millennium Footbridge across the Thames between St. Paul's Cathedral and Tate Modern, which came to light when the first pedestrians crossed it, was treated with broad good humor.

While it might be considered desirable that art should involve the sensation of taking a risk, then, any actual risk must of course be minimized in the same way that it is for other facilities open to the public. Ignoring Höller's sociologist, the administration at Tate Modern was therefore keen to stress not the feelings of panic that *Test Site* might have induced, but the stringency of the safety checks that the work underwent. And even the artist himself felt driven to point out of his slides: "They are much safer than stairs; stairs are quite dangerous."

But perhaps this is to overemphasize the need for safety. Many people seem unexpectedly prepared to make allowances for art. Even at the time of the umbrellas tragedy, Californians seemed to sympathize as much with the artwork and its creator as with the human victims. "It was just an accident of nature. Why should it ruin everything for the rest of us?" responded one visitor to the site, a retired construction worker, when questioned by the *San Francisco Chronicle*.

Some time later, the lifestyle supplement of the *Los Angeles Daily Times* headlined: "Umbrellas: a year later shock of deaths wearing off." The article discussed Christo fans' plans to commemorate the short-lived art installation and took the unusual step of quoting a professor of art on the matter of risk assessment. "People get killed in building roads and bridges and no one thinks about it. Art should be judged on the same scale," he felt.

In fact art *is* judged on the same scale because accidents involving art are not so numerous that they warrant a category of their own. Because he was at work, the Japanese crane operator on Christo's umbrellas is regarded as just another construction industry fatality. The death in 1983 of a studio worker of the New York sculptor Beverly Pepper, who was crushed when one of the artist's iron sculptures fell on her, was likewise an industrial accident because it happened in the workplace.

When members of the public are the victims, the situation becomes less clear. The U.S. National Safety Council lists deaths by 117 accidental means, including "bitten or crushed by reptiles." Christo's Californian victim would simply have been "struck by or striking against object." The distinctions are even hazier in the UK, where the Health and Safety Executive records the 200 or so fatalities a year at work in gory detail, but the 12,000 or so other accidental deaths each year simply by where they happen, mostly "at home" or "elsewhere," rather than by cause.

Two more conclusions may be drawn from these examples. The first is that art simply becomes more dangerous as its scale expands—when it becomes very large or massive, when it requires numerous crew members to set it up and take it down, when it covers large areas, when those spaces are unrestricted to the public, when those spaces present wild or unfamiliar terrain, and so on. In the case of Christo's umbrellas, two unconnected deaths make the art look dangerous indeed. But, set against the scale of the work, the coincidence is less remarkable.

The final reason for the apparent growing danger posed by art is its growing popularity. Half a million people were reported to have seen Christo's umbrellas in Ibaraki. Seven thousand turned up on one day in California. Each week, 44,000 rides were taken on Höller's slides at Tate Modern. If nobody visited these works, there would be fewer human injuries.

Cheers!

"Students mark 21st birthdays with 'extreme' drinking binges"
—*USA Today*

"Effort to curb binge drinking in college falls short" and "21st birthday booze ritual gains popularity" are typical of the recent headlines in American media. The problem—known as the "English disease" but almost as prevalent in the United States and Canada—seems to affect adults, youths, and even children. One journalist describes pub closing time in an English town: "When the pubs shut, the drinking tribes charge out like wounded bulls, piss in the alleyways, wrestle with the rubbish bags, smash bottles on the pavement, break the occasional shop window and do a lot of braying."[1] Other countries have heavy drinkers, but the gratuitous violence and vandalism are not on the same scale—though certain American demographics certainly try their best.

The tendency for people to drink more on the occasions that they do drink—"binge drinking"—is higher now in virtually every country compared to a decade ago. According to the International Center for Alcohol Policies, a not-for-profit organization funded by leading alcohol producers in America, there are diverse definitions of binge drinking, more formally referred to as heavy episodic drinking.[2] One of the increasingly common thresholds is for men drinking five or more drinks or for women drinking four or more, on one occasion. Canada defines binge drinking as the consumption of eight drinks within the same day, while Sweden considers half a bottle of spirits or two bottles of wine on the same occasion to constitute a binge. Such serious drinking usually occurs in large groups.

There are some occasions, and some cultures, where heavy drinking is accepted: for example the rites of passage into adulthood (Japan and the Pacific Islands); certain university cultures (in the U.S. and Canada); at sporting events; and for exceptional celebrations beyond conventional behavior, such as weddings, New Year's Eve, and Mardi Gras celebrations. In Europe, binge drinking is most prevalent in northern Europe, notably Scandinavia, and least common in the southern part of the continent, in Italy, France, and other Mediterranean countries.

But it is Britain and Ireland where the culture of drinking has changed most for the worse in recent years—and has a grim reputation. David Ginola, the talented French soccer player of the 1990s, said that he would not bring up his children in England because "I don't want my daughter to be an Englishwoman." He observed that across the country he saw women trying to keep up with men, drink for drink, usually concluding with unsavory behavior such as "vomiting in the streets."[3]

Survey data seem to support Ginola's view. A sample of British teenagers aged between 12 and 17 suggests that 5 per-cent have been so drunk that they have had their stomachs pumped and a further 13 percent admitted suffering from such a bad hangover that they had played truant from school.[4] But the U.S. isn't doing much better on this front: The 2007 Youth Risk Behavior Survey found that 11 percent of high school students drove after drinking alcohol, and 26 percent rode with a driver who had been drinking.[5] The problem, according to the charity Alcohol Concern, is not that more young people are drinking alcohol than before but that those who do drink, drink more. As official figures show that over 8 out of 10 American high school students have drunk alcohol, there is little scope for the participation rates to rise.

Heavy drinking is associated with a series of negative outcomes. Some of these are medical in their nature, leading one EU-funded report to say that "Alcohol is public health enemy number three, behind only tobacco and high blood pressure, and ahead of obesity, lack of exercise, or illicit drugs."[6] Dangers of alcohol include an increased risk of stroke and other cardiovascular problems, neurological damage, and adverse effects on the health of the fetus of a pregnant woman. Drinking can, of course, also damage the liver—one British newspaper noted a 37 percent rise in the number of drinkers dying from alcoholic liver disease in the last 5 years, with hospital admissions for the condition doubling in a decade to over

35,000 a year. Alcohol can also lead to poisoning. One article told us of three young children in Montana—ages 9, 11, and 12—who died of alcohol poisoning.

However, it is the adverse social consequences that generally attract more attention in the media. A National Institute on Drug Abuse report estimated that the tangible social costs of alcohol in the U.S. were around $57.2 billion—nearly equal to the $66.8 billion cost shouldered by the alcohol abusers themselves and their families.[7] These include: unemployment, absenteeism, traffic accident damage, criminal damage, the cost of police, courts, and prisons, and the medical bills. An EU report estimated that the costs in the workplace, along with those due to crime and traffic accidents, were in total at least four times as great as the health-care costs. The addition of the more intangible costs, such as the loss of healthy life, impaired professional and academic performance, anti-social behavior (and hooliganism at sporting events), and relationship difficulties, including the suffering of domestic violence victims, would add considerably to the cost.

Young people are, for a variety of reasons, at increased risk of harm due to excessive drinking. Many risks stem from young people's relative inexperience with alcohol consumption and a greater tendency toward risk-taking, but still-growing young people are also more susceptible to brain and other physiological damage. Young drinkers also have a greater chance of alcohol dependency later in life—or so we are told. The *New York Times* ran a story, "The grim neurology of teenage drinking," which explained that children drinking heavily at age 14 or younger have a 50 percent chance of becoming an alcoholic—a risk roughly 5 times greater than for the general population.[8] The problem is that people who start drinking that early are often very different in a whole host of ways from those who don't. They are more likely to have at least one parent with alcoholism—itself presenting a 40 percent risk of alcoholism for the child—and are more likely to come from chaotic homes and to have suffered from child abuse.

According to one American newspaper article, underage drinkers find it easy to get hold of alcohol via the Internet. One article told us that "one in 10 teenagers have an under-aged friend who has ordered beer, wine, or liquor over the Internet."[9] It also told us that more than one-third thought they could easily do it and nearly a half thought they wouldn't get caught. All of this sounds pretty scary, but closer analysis suggests that the survey, which was paid for by the Wine and Spirits Wholesalers of America, a group that competes with direct sellers of alcohol, perhaps tells another message that suited the sponsors less well and would have had less media impact.[10] Despite most teenagers using the Internet daily and often spending many hours online, and despite almost half believing they could buy alcohol online, only 2 percent had ever done so. This is not a very large proportion given almost half of those surveyed were between 18 and 21 of age.

Heavy drinking is clearly a problem that we could live without, and the press and politicians regularly debate the available policy options. In 2001, New York City Mayor, Rudi Giuliani—fed up with booze-related violence and vandalism—tried to ban alcohol from all New York City street fairs. Higher taxes, especially on the drinks favored by young people, could also play a part in reducing consumption. One Scottish community is hoping to reduce crime by encouraging local shops to ban the sale of alcohol after 7 p.m. Others advocate that premises only be granted a licence to sell alcohol if there is plenty of seating, on the grounds that bars with standing room only encourage more rapid consumption of alcohol. But pessimists argue that the problem runs deeper than drinking regulations and is rooted in a working-class tradition of loutish youths "moving in packs"—groups that are at their most visible when traveling.

Treating the English disease is definitely proving difficult. In 2005 a British government initiative removed the traditional eleven o'clock closing time of pubs, allowing them to apply for licences to stay open later. They argued that this was "the European way" and that by easing pressure of last orders, binging would be reduced.

Those against the change said that it would merely prolong the misery and cause more damage. One year on from the change, the impact seemed not to have been significant one way or the other—although the government, which has learned to manage the news flow, did leak to the press prior to the publication of the figures that the number of drunk and disorderly people decreased in the period immediately after the change in the law.[11]

In the States, the focus seems to be on saving victims' lives. Due in part to the support of organizations like Mothers Against Drunk Driving, law enforcement measures such as sobriety checkpoints and ignition interlocks (breathalyzers built into cars' ignitions) have gained legislative popularity. An EU report similarly advocated more extensive random breath testing of drivers, higher taxes on alcohol, and shorter opening hours for the sale of alcohol, as well as restricting the extent and content of alcohol advertising. It also wanted tougher measures aimed at young adults. There is no doubt that one constraint on government's pursuit of such policies is the size of the developed world's drinks industry, which takes advantage of its association with prestige products and even national identity to foster close contacts with ministers. There are roughly three-quarters of a million jobs in Europe's drinks production industry, with many others indirectly linked to the business working in shops and bars, and it produces billions in tax revenues annually. France and Italy together account for over half of the world's wine exports, the UK and France account for over half of the world's spirits exports, and the Netherlands and Germany account for over one-third of the world's beer exports. Governments end up treading very carefully.

Social attitudes might change. In the nineteenth century, European elites were faced with a situation of urban squalor that included unprecedented public drunkenness. This led to the growth of temperance movements across Europe. Initially they were focused on the idea of drinking in moderation and only later moved on to the idea of prohibition. Finland, Iceland, and Russia were among the countries that adopted some form of prohibition, while other countries allowed individual areas to vote on prohibition. Might we again see bans on alcohol? The City of York Council, for one, has drawn up proposals for an Alcohol Exclusion Zone in two parts of the city in response to pleas from residents who have experienced alcohol-related disorder.

While there is no doubt that consumption of large quantities of alcohol is damaging for health, there is some evidence—especially popular among wine drinkers—that modest quantities of alcohol can give beneficial effects. A study in the *Lancet* over twenty-five years ago compared the death rates of men in their fifties and sixties from heart disease in a number of countries. It found the highest death rates were in traditional beer- and spirits-drinking countries, while France had the lowest number of deaths and the highest wine consumption. A Danish study of 24,000 people also found that drinkers of wine, as opposed to other forms of alcohol, benefited from an overall reduction in deaths from all causes. But the village of Gers in southwest France seems to provide the conclusive evidence that wine is good for you—it has double the national average of men aged ninety or more. This is put down to the wines being very rich in procyanidin, which somehow seems to counter the otherwise negative effect on life expectancy of the French diet, which is typically high in saturated fats.

This positive view of moderate drinking is not new. The Greek physician Hippocrates was using wine as an antiseptic, diuretic, and sedative 2,400 years ago. Louis Pasteur, no less, said in the mid-nineteenth century, "Wine is the healthiest and most hygienic of drinks," and in an era predating modern manufacturing and sterilizing processes, wine provided products with a stability and cleanliness that water could not. Even before Pasteur, a serving of wine each day on the convict trips from England to Australia contributed to a reduction in malnutrition and scurvy during transit. So successful was this that some of the British doctors involved started their own vineyards in Australia.

For now, binge drinking in the U.S. appears to have plateaued, probably held in place by the college age group, for whom binge drinking seems a staple of life despite any measures taken against it. Meanwhile, those of us who have long since left teenage years (and binge drinking) behind will hope that none of the measures that could be introduced to restrain the abuse of alcohol would affect the—essentially immeasurable—pleasures associated with alcohol. The trouble is that some old people cannot be trusted. "Senior-citizen drinking problem labeled 'invisible epidemic'" was the headline of the story saying that one in six Americans over 60 is overdependent on alcohol. There seems to be no easy solution.

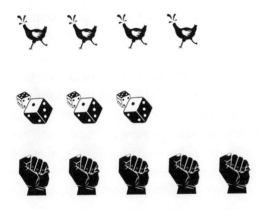

The Death of Cinema

"Theater could face final act"
—*The Washington Post*

The arrival of digital videos and cameras, now typically available in cell phones, and Internet hosting sites, means that film-making is no longer the preserve of a few Hollywood barons. The phenomenal success of YouTube, the site launched only in 2005, where people can watch and share original videos, exemplifies the way creative possibilities have opened up for the masses. But does it mean the death of the cinema?

We will certainly never be short of recorded moving images. In 2000 alone, it is estimated that 1.5 billion hours of moving images were created, roughly 200,000 hours of film every hour. That represents a doubling compared to a decade before, but nothing compared to the forecasts for the years ahead. Within twenty years we could be creating 100 billion hours of moving images every year—equivalent to over 10 million hours of film for every hour of the day.[1] Such a volume of material presents a range of difficulties, not least choosing what to watch and what to archive and save for the future.

Meanwhile, the increasing availability of DVDs, often at very low prices, and improvements in home entertainment are contributing to the fall in cinema attendance. This in turn is putting considerable financial pressure on the entertainment market and prompting much head-scratching about how to maintain revenue streams going forward. In the same way that digital technology has changed the production of film, for example in editing and the creation of special effects, it will surely have an impact too on distribution, projection, and home delivery. This could be viewed as a positive development, not least because the lower costs allow new, independent directors to produce films. But where will we watch the new films?

The movie business has already been through—and survived—a number of transitions during its relatively short life. The early publicly shown movies—silent and black and white, probably part of traveling exhibits or acts in vaudeville programs, often only a minute long, perhaps showing a single scene, authentic or staged, shown largely to working-class audiences—would be unrecognizable to today's cinema-goers.

The First World War wrought many changes, notably dealing a devastating blow to the European film industry and giving Americans—with the rise of Hollywood—the chance to earn the dominance in the mainstream industry, a dominance that it still has today despite many other countries and cultures having more or less thriving genres at various times. The hundred-year history of cinema shows the coming and going of many styles, techniques, and fashions, and no shortage of people who say that things just aren't what they used to be.

The economics too have changed. Cinema-going in the U.S. reached its peak in the 1940s, when it was estimated that Americans spent nearly one-quarter of the money available for recreation on going to see films, compared to no more than one-fiftieth, 2 percent, today. The widely accepted principal explanation for this decline was the mass arrival of televisions—between 1950 and 1955 the number of televisions in America increased eightfold to over 30 million. Although it is likely that increased urban sprawl (most cinemas were in urban areas) and anti-trust action (studios were no longer permitted to own theater chains) also played a part.

But the long-term decline in the number of admissions stopped and rose for a couple of decades, due mainly to the arrival of the multiplex screens, which offer both more films and more viewing times for the same film. The number of screens in America increased by over 50 percent to 37,000 during the 1990s, with a similar rise in the UK a little later, as multiplexes were rolled out. But for all the positive medium-term trends, the very recent past has seen declines in admissions again, reigniting concerns about the future. In the U.S., admissions peaked in 2002 at over 1.6 billion and have since declined to 1.4 billion, returning admissions to the 1997 level. The fall in Britain over that period was smaller—though 7 percent through to 2007—and much smaller across Europe. Such declines are noteworthy but hardly amount to a death. In any case, attendances continue to rise in the Asia-Pacific region—up by 22 percent in the last four years. The strength in the Asian market is encouraging for the business as it comprises around 60 percent of global admissions.

According to the Motion Picture Association of America, the number of films released for a week or more during 2007 remained on par with the 603 released in 2006 and would seem to suggest that the industry is healthy, but the top twenty films drive the business as they still account for roughly half of box-office takings.

Even so, it is clear that a greater variety of films is available than was the case a decade or two ago.

Paradoxically, one reason to be optimistic about cinema attendance is that people in many countries do not go to the cinema very often, leaving plenty of room for improvement. Americans on average visit theaters five times a year, but Britons, French, and Spanish visit the cinema no more than three times a year on average—and that is roughly twice as frequent as Italians and Germans. Plenty of youngsters go to the cinema frequently, but attendance drops off as we age—a trend that the industry could try to reverse.

The battle fought by cinemas in the last two decades has been against the VCR and DVD. While it survived the VCR, the latest decline could be linked to the increase in DVD household penetration. The number of American households owning a DVD player has more than trebled from 25 million in 2001, an increase from under one-quarter of TV-owning households to over three-quarters. Over the same period, the average price of a DVD player halved, and the number of titles available trebled. Cinema admissions per head in America have fallen from a recent peak of 5.7 a year in 2002 to 4.8, with larger and better home TV screens, broadband Internet options, more accessible illegal downloads, and pirated DVDs putting pressure on numbers. On the other hand, research suggests that the households who own more sophisticated home entertainment technology tend to visit the movies more frequently than lower-tech households.

In the UK, the market for the rental of videos and DVDs has declined by over one-quarter from the peak in 2000, but that has been more than compensated for by an increase in the retail market. The total sales, of 229 million videos and DVDs in 2006, was more than double the sales seen in the late 1990s and reflected the falling price of DVDs as people have replaced the old-style video collections with the new media. The UK especially has also seen a dramatic rise in the occurrence of newspapers giving away discs—roughly 130 million units were given away in 2005 alone, with the number given away in the first quarter of 2006 being broadly the same number as retailers sold through traditional channels.

The next big advance for the cinema-goers will be the arrival of digital cinema, which will not only reduce costs for distributors and help to control piracy, but, it is hoped, will also maintain the viewers' experience gap with regard to the developing home cinema and high-definition TV market. This might well amount to the "death of film," if not the cinema. Cinemas are now rolling out the new technology, and more than 20,000 "3D" screens are expected worldwide by 2010. It sounds like an exciting enough prospect to keep drawing crowds, and, in any case, the cinema will still be a good place to take the children on a rainy day during the school holidays. Technically at least, such digitalization narrows the gap between television and film, opening the way for events such as concerts and sports to be shown in the cinema. As these events already draw much larger audiences than the most successful Hollywood film, the cinema, perhaps renamed and remodeled, looks likely to survive into the future.

4. Social Policy

We are always worried about money—pensions, debt, and housing are the big-ticket items at the top of the list. Get these wrong, and your life can be ruined. A credit crunch, house price collapse, or commodity price hike could also spell trouble. Government economic and social policies, for example on immigration, can affect us too and so can being in the wrong place at the wrong time—if you happen to be involved in a traffic accident.

Golden Oldies' Time Bomb

"Can we outgrow our social security problem?"
—*U.S. News & World Report*

A generation ago, the life pattern for most of the working masses in the developed world was simple: they worked until they were sixty-five (retirement age) and then they died. But increased life expectancy has changed all that, leaving people in the position where plenty have to plan for around two decades of life in retirement. This is a big change since the first universal social security in the U.S. was introduced nearly a century ago, in 1935. That applied only to anyone aged over sixty-five—almost a decade above life expectancy at the time—on a means-tested basis.

The phrase "pensions time bomb" has been coined to refer to the predicament we face. A growing proportion of elderly to care for, as explained in Chapter 1, with a smaller proportion of workers paying taxes to fund that care means that government will be unable to afford generous payouts in the future. The reluctance, and in the case of the lower earners, the inability, to make private provision for pensions makes the situation worse. Company pension schemes are also under threat. The system in many countries is now complex, unwieldy, and in crisis.

The institutional differences between countries coupled with decades of widely diverging savings behavior make it very difficult to compare the state of the pensions provision in different countries. Many also suffer from a lack of decent statistics—and decent forecasts—setting out the state of play and the scale of any potential problem. Figures from state administrative systems and private savings schemes are often incomplete, pay-as-you-go systems have no statistics, and surveys are notoriously poor at collecting pension data as respondents rarely know the answers to questions.

Pension funds play a significant contribution in only a minority of developed countries. Four countries—the Netherlands, Iceland, Switzerland, and the U.S.—have pension fund assets that are greater in value than the annual output of their economies. Others, including Canada, Ireland, Australia, and the UK, have assets of between one-half and one-third of national income. In terms of absolute value, American pension funds dwarf those in any other country, accounting for two-thirds of the global total. The UK's assets are the second-largest, almost equal to the combined value of the third- and fourth-placed countries, Japan and the Netherlands. The dominant position of the Anglo-Saxon countries—Canada and Australia also have large pension funds—reflects the maturity of the private occupational pension plans started decades ago.

The slump in world equity markets between 2000 and 2003 caused the value of these assets to fall until the rally in the markets saw the value in nominal terms increase substantially (to $18,000 billion in 2005 from $13,000 billion in 2001). This rise in the value of assets and the more recent weakness means that across the industrialized nations as a whole the value of the funds in relation to national income is unchanged from what it was at the beginning of the decade.

Many countries, including Spain, France, and Italy, have followed a different model, where public pensions play a dominant role in the old-age retirement system. In these cases, pension fund assets amount to less than 10 percent of national income. Many countries with small or immature private pension schemes have introduced policies to enhance them in the last decade. Where this has happened, for example in the Czech Republic, Hungary, and Poland, asset values have grown quickly, although they are still small.

In addition to private pension funds and life insurance assets, several countries have accumulated assets in their so-called national pension reserve funds (funds set aside by otherwise pay-as-you-go systems in preparation for the rising financial costs resulting from the predicted aging of populations [and the decline in the proportion of the population that is working that can finance the pensions through their taxes] in the next few decades).

Countries need to address these issues because the levels of public spending on social benefits are already high in many cases. Four European countries—Sweden, Denmark, France, and Germany—spend an amount equivalent to over 30 percent of their national income on social benefits and will find the costs increasingly hard to meet as the demographics deteriorate.

The rise in national pension reserve funds is one of the responses from governments, which are becoming increasingly aware of the financial problems that many of their citizens will face in retirement as funding pensions from pay-as-you-go schemes become more impractical. There are various ways that governments can reform their pensions systems, including improving the regulatory environment to reassure savers that their money will be safe and creating a tax regime to incentivize saving. Governments have also introduced policies to encourage people to stay in the workforce longer, and since the year 2000, age discrimination in employment has been outlawed across the European Union (though under the Age Discrmination in Employment Act, it has been illegal in the U.S. since 1967). Many of these policies are new, and it will take many years to see an impact on the official statistics. As it stands, average retirement ages in Europe range from 63 in Sweden to 56 in Slovenia. The gap between life expectancy and the average exit age from the labor force varies between 21 years in France and 8 years in Latvia, with over half of the countries bunched between 15 and 18 years. In the U.S., the gap is currently around 13 years.

The trends strongly suggest that many—perhaps a majority of—people in developed countries are unlikely to have a financially comfortable retirement as they are either choosing not to save or are on incomes so low that saving is not a realistic option. It is no surprise that governments find it very difficult to introduce mandatory savings schemes on a scale that would be likely to have a material impact on future pension provision. Company pension schemes, upon which many employees are depending, are also unlikely to be the solution. One Scottish newspaper article, "Pensions crisis is much worse than firms say," quoted the president of the Faculty of Actuaries, Scotland's most senior actuary, as saying: "We are kidding ourselves over the security of pensions." He predicted that more pension schemes will collapse and pensions be lost, because companies rely upon "wholly inadequate" yardsticks to measure solvency.[1]

AON Consulting, part of the insurance and risk management company, produces an annual benchmarking study of the European pensions systems, assessing which country is likely to find itself under the greatest pressure to make material changes to its pension system in the medium to long term. The study analyzes each country according to its demography, adequacy of the state pension, affordability and sustainability of the state pension, and the availability of company pensions. The 2006 study put Denmark in the most favorable position and Belgium in the least favorable. The top countries, including Denmark, Estonia, Ireland, Latvia, the Netherlands, and the UK, combined favorable demographics with reasonable private pensions. The state pensions might be poor but at least they are affordable. France, among the lowest-performing countries, combines a high life expectancy, one of the lowest retirement ages, decent state pensions, and low private provision, similar to the situation in Belgium.[2]

The U.S. is one country that has at least conducted an analysis of the problem and identified possible solutions, even if the appropriate policies have not yet been fully adopted. The Commission to Strengthen Social Security was set up by president George W. Bush in 2001 to investigate ways of restoring fiscal soundness to the social security system. It produced three models for modifying the program, the tenets of which were advancing personal funding (via the wealthier members of society), increasing the benefits to lower-wage workers, and making the system self-sustainable. Two main priorities of the proposed models were to privatize the system and not to increase the level of taxation. Unsurprisingly, the report was criticized for a lack of progressive thinking. Needless to say, social security reform in the States is currently pretty stagnant.

Another suggestion has been to encourage delayed retirement, which has been an unpopular proposition, in part due to suspicions that delayed retirement would lead to an increase in the unemployment rate, which is already quite high.

Other countries have introduced a variety of initiatives that are likely to help: Turkey has raised its retirement age; Finland has changed the formula used to calculate benefits so that they are based on earnings throughout the working life rather than a specific shorter period of relatively high earnings; Mexican legislation has increased competition among pension fund managers and cut the costs of running such schemes; and Italy has legislated to boost the growth of private pensions. In all, bar a few developed countries, the proportion of defined benefit plans, where the pension payable is related to income earned, is decreasing while the proportion of defined contribution plans, where the pension is related to the value of savings and investments made, is increasing. Schemes of the former type are generally more generous.

Despite repeated attempts by the government to increase privatization of pensions, the population does not seem to have changed its habits. It seems that paying off a mortgage, repaying a student loan, or just spending money are far more appealing options than making a decent provision for retirement. Perhaps people have been turned off by the idea of saving by the stories of individual pension fund scandals. A report in 2006 on the City of San Diego's pension fund scandal told of "years of reckless and wrongful mismanagement" and "non transparency, obfuscation, and denial of fiscal reality."[3] The report drew analogies with what happened at Enron and HealthSouth in the U.S. Such stories go back many years including the plundering of the UK's Mirror Group pension fund by Robert Maxwell, which was discovered after his death in 1991. The pensions mis-selling crisis in the UK in the late 1980s is also fresh in the memory.

The gradual decline in the percentage of the British workforce that has any current pension provision above that provided by the state has continued, with the decline in participation rates among private-sector workers continuing at a significant pace—from 10.5 million to 9 million in the last eight years. The percentage of employers making any pension provision for their employees declined from 52 percent in 2003 to 44 percent in 2005. Some people have no doubt been hoping to be bailed out by their investment in property, but house price weakness means that only a minority will be saved, and many of those who have done well in the property market have probably also made some provision for a private pension.

The prospect of poverty in old age remains very real for many people. Most countries have to face up to significant policy changes or come to terms with the consequent growing inequalities in society.

Credit Crunch

"The illusion of cheap borrowing"
—*The Wall Street Journal*

During 2008, at the time of writing, the most talked about threat to the stability of our lives was economic recession. A combination of factors but mainly the credit crunch—the increased difficulty of getting credit following the demise of the sub-prime mortgage market in the U.S. in 2007—falling house prices, and higher commodity (mainly oil), and food prices were threatening to derail the world's economy.

Clearly the financial golden age was coming to an end, but the question was how serious the slowdown would be. As recession loomed, the economic fundamentals were poor, especially in the U.S. where the bank reserves were beginning to flounder.

But the real problem was the *fear* of what might happen. Nouriel Roubini, the nobel laureate for economics, predicted a recession in 2006 and was ridiculed for it, but he was right. Interest rates were a bit higher, the housing market was weakening and banks were reluctant to lend, but if recession was deep there was every chance that it would be noted as one that was brought upon us by fear rather than fundamentals.

At the heart of the problem was borrowing. Easy credit from banks at a time of low interest rates and prolonged growth had led to a binge on a scale never seen before. Americans were the biggest borrowers and the most profligate spenders—and the figures were frightening. In 2005, Americans took out $750 billion worth of equity from their homes (an increase from the $106 billion they extracted in 1996)—most of which were bought on borrowed money—and spent the majority of it. Total U.S. household debt in 2006 was $12.8 trillion, or twice the amount of household debt in 1999. The average amount borrowed on credit cards in 2005 stood at $12,400 per person. This might have dire outcomes for the feckless families involved, but, given the tighter credit regimes from banks since the end of 2007, the problem would push the economy into severe recession.

The excesses became a real problem in late 2007 when the credit crunch started. Suddenly it was harder to refinance borrowing and lenders took a less lenient line with those who could no longer repay.

The young, who are doing more than their fair share of the borrowing, have been described as the iPod generation—insecure, pressured, over-taxed, and debt-ridden.[1] Traditionally the young take out debt to buy their first property and the middle-aged repay the mortgage and save for old age. The elderly run down their savings and sell their assets. But that life cycle is being distorted by several factors that encourage the young and middle-aged to borrow more than they have in the past. The large inflation-adjusted increases in house prices in recent years mean that proportionately more now has to be borrowed to get on the property ladder. Today's twenty-somethings are the first generation of graduates to start their working life in debt. A survey of students showed that, while one in eight most feared a terrorist attack, one-third most feared going deeply into debt and another third feared unemployment.[2] Nearly a half of those surveyed expected to graduate with $10,000 or more in college loans, with one in five saying they would have more than $20,000 to pay off.

The "buy now, pay later" culture has taken a firm grip on consumers in many countries. And all of the evidence suggests that those who want to have children are having to deal with costs increasing rapidly in real (inflation-adjusted) terms.

This love of credit is not evenly spread across all countries. An Organization for Economic Co-operation and Development study of fifteen industrialized countries showed that Britain, the Netherlands, and Denmark actually had total household debt that was greater than national income.[3] Italy and Finland were at the other end of the scale, with debt levels less than 50 percent of GDP. Except in Japan, debt has increased, in most cases significantly over the last decade, but Britain is the dominant borrower in western Europe.

The U.S., the Netherlands, and the UK are three countries where over four out of ten households have mortgage debt—in Italy, Germany, and Spain, the proportion is below one-quarter. There were 5 countries in the OECD study where at least 4 out of 10 households have debts other than their mortgage—New Zealand, the U.S., Canada, Sweden, and the UK. Italy, Germany, and Spain were again the countries where the proportion of households with debt was the lowest. (Debt can also be measured in relation to annual disposable income and net wealth. While the pecking order of countries will change under different measures, the main differences between countries tend to be maintained.)

One report claimed that overspending Britons were responsible for a third of all unsecured debt, namely that excluding mortgage debt, in western Europe. The £3,000 (around $4,000) per person owed is roughly double the average elsewhere in Europe. Some experts suggested that Britain was approaching its debt "saturation point" and that lenders would have to focus on other markets to seek high returns.

There is nothing wrong with borrowing in itself. Indeed the redistribution of capital from those who wish to save, and earn interest on, their surpluses to those who are willing to pay to use the money is a driver for economic growth. Borrowing, repaying debt, and then saving is a common life-cycle activity for most people in developed countries. And the increase in borrowing should be expected following the progressive removal of credit rationing with the financial deregulation that is currently taking place in many countries. Lower interest rates, both in nominal and real terms, have also boosted demand for credit. That said, the size of the total personal debt mountain looked worryingly large as the credit crunch took hold.

The high debt figures are already a matter of concern for a significant minority of the population, as is nowhere more evident than from looking at the number of debt help services that have sprung up. Mortgage delinquency, or default, rates have edged up in most countries from recent low points. Sales numbers for certain products make grim reading too—Pfizer's antidepressant Zoloft saw a 200 percent increase in sales in 2008, and alcohol sales are spiking similarly.[4]

And the problems started happening when there was a benign economic environment of low interest rates and relatively high employment. Those in debt become particularly vulnerable when the economy weakens, at which point many more households could be pushed over the edge into financial crisis. The international Oranisation for Economic Co-operation and Development believes that households in aggregate have become more vulnerable to adverse shifts in the economy.[5] The FSA has warned of various risk events that could affect financial markets and the ability of people to service their debts, namely a global pandemic, terrorism, financial system crisis, and a significant fall in house prices or major corporate bankruptcy. Two other worries were ushered in by the "credit crunch" of 2007, which led to higher interest rates for borrowers with lower credit ratings. The British government's decade-long claim to have put an end to "boom and bust," itself the type of rhetoric that over-encourages borrowing, has proved to be wrong. At the time of writing, most economies are already in—or on the brink of—recession.

For all the doom and gloom, however, the situation might not be as bad as it seems. Households' net wealth has increased, mostly reflecting a sharp appreciation of property values and an increase in home ownership rates, and incomes have risen. It also seems to be the case that most indebtedness across the household sector is held by higher-income households that are generally better able to manage it. This needs to be taken with a pinch of salt, however, as it might be those on lower to middle incomes who have difficulty meeting their repayments in a crisis even though their debt levels are lower—those on lower incomes generally have less in the way of reserves to draw on in a time of crisis and, with inferior credit ratings, might well have access to less flexible borrowing schemes and be subjected to higher interest rates.

There is little sign to date that enough households are having difficulties meeting their financial obligations to lenders to harm the broader economy, but economists keep trimming their growth forecasts, lengthening the expected duration of the recession. It does seem, however, that an increasing number of families are engaging in what has been described as "revolving arrears," that is, shifting debt to where it can be most easily and most cheaply handled, for example the utility companies. And utilities companies are responding in the only way they know how: Pennsylvania's PPL Core reported a 78 percent increase in power shut-offs in the first three quarters of 2008 compared with the same time in 2007.

Even if a serious crisis is averted, borrowing has other consequences. Higher debt does serve to shift financial obligations between generations. More borrowing and less saving now by today's workers will leave them with lower pensions and less equity than they expect in the future, and quite probably a more financially stretching retirement. It will also mean that there is less money to pass on to the next generation—properties might have risen in value but so have the debts set against them.

The Housing Bubble

"What real estate bubble?"
—Time

Panic about the housing market in recent years has been widespread and came in two particular forms—first, there was the worry that prices were so high that people who were not already homeowners would be permanently excluded from the market; second, there was the worry of the economic consequences of a material fall in prices should the so-called "bubble" burst—which it did. The large increase in prices over the last decade or more has left us with a nasty hangover—the only uncertainty being the outcome of the social and economic crisis we find ourselves in.

At the time of writing, housing markets around the world have slowed and some—notably the Irish, American, Spanish, and British—have seen hefty price falls. The newspaper headlines would have us believe that markets are in freefall. But they are not—yet. Even in countries where the aggregate position is negative, the performance of localities is diverse—some have lost the last few months of rises while others have lost years of rises. The falls are very painful for some families and are causing economic problems in some countries but for most home owners the markdowns are just a paper loss. There is little doubt that the hefty house price inflation of recent years, barring a major fall, will lead to greater social division as the rich get richer and it becomes harder for first-time buyers to enter the market. "House price explosion" and "Property ladder too high for 17 million" are typical of the large-print, panic-inducing headlines that regularly appeared over years on our front pages as markets kept rising into 2007. There are particular concerns for key workers and young people. How can teachers or nurses be expected to buy property in the highly priced urban areas in which they work? Commuting is often not an option for those on low pay or working anti-social hours. Labor mobility is also curtailed as people find it impossible to move from areas of low house prices to areas of higher house prices. Alternatively, a price correction—shall we call it a crash?—could leave the economy in tatters, given the debt mountain and speculation (especially in the buy-to-let market) that has supported the rise in house prices.

House prices have been rising faster than general inflation and earnings in most countries for some years. With the exception of Germany and Japan, advanced countries have been in the grip of a housing boom since the mid-1990s. The Paris-based think tank, the Organization for Economic Cooperation and Development, has described this boom as unprecedented in its steepness, durability and geographical breadth.[1] The *Economist* magazine described the worldwide rise in house prices as "the biggest bubble in history."

The most notable growth was in Ireland, where house prices more than trebled after allowing for inflation since 1992. Prices doubled in the U.S. in the last decade. It is daft money. In 2006, Donald Trump's Palm Beach mansion and Tommy Hilfiger's mountain home in Nevada were each put on the market for around $120 million, a figure surpassed only weeks later when a Saudi Arabian prince's ranch near Aspen, Colorado, was offered for $135 million. Trailers on the Californian coast sold for more than $1 million. A villa in the south of France was sold for €500 million (roughly $630 million).

In 2007, the world saw several markets pause for thought and in some cases declines occurred in 2008, but where are they heading? The positive housing market factors are common to most of those areas that have experienced above-inflation increases: growing demand for property (from families, immigrants and speculative investors), building land becoming increasingly hard to find especially in the more desirable areas, and interest rates and unemployment at historically low levels. The increase in immigration has contributed to the price rise as well. The government never adjusted the housing program to take account of the millions of people who have immigrated to the U.S. since 2000.

The rise in property prices has given homeowners considerable equity with which to buy additional properties, often on a buy-to-let basis, inflating prices further. And until 2007, at least, there has been no shortage of imaginative, alluring, and generous financing deals from the mortgage lenders—zero down payments, low starter rates, loans higher than the valuation of the property, flexible payments, and even "stated income" or "self-certification" applications, in which the borrower is left to use his own imagination to describe his financial circumstances.

This was fantastic news for homeowners. As they saw the value of their assets rise, it made them feel more financially secure about the future. There are many people who for some years "earned" more money from the rise in the value of their property than they did from their work. The Office of Federal Housing Enterprise Oversight reports that in 2005, house prices shot up 12.5 percent while the prices of other goods rose only 3.1 percent.[2] The newspapers, and presumably most of their readers, loved stories telling us about the massive increase in the number of people living in properties valued at over $1 million.

On the downside, housing unaffordability—measured in many different ways but essentially the cost of property relative to people's incomes—got to an all-time high in many countries, reducing the supply of buyers. In the U.S., the price increases were so universal that housing prices rose even in places that had been recently affected by Hurricane Katrina.

There are pockets of unaffordable housing around the world. One study calculated the median house price to median household income ratio for a number of cities and regions in a handful of Anglo-Saxon countries. It described anywhere with a median multiple of over 5 as being "severely unaffordable"; a score of 3 was described as "affordable." Three locations—Los Angeles, San Diego, and Honolulu—had a median multiple score of over 10. Another 20 locations, about half of which were American, achieved a score of over 6. These included San Francisco, Miami, Sydney, New York, and London.[3] Roughly one-quarter of the locations studied were deemed as affordable and another quarter as "moderately unaffordable," leaving the other half to be severely or seriously unaffordable. Those deemed affordable in the U.S. and Canada included Houston,

Atlanta, Dallas, and Quebec City. None of the British locations qualified as affordable.

The study described the last decade's price escalation as "unprecedented" and said that it was "a matter for concern." The locations studied in Australia, the UK, New Zealand, and Ireland (Dublin) had an average score of around 6, compared to 4½ in the U.S. and just under 4 in Canada. The high scores were all the more surprising as most of the locations had ratios of around 3 as recently as a decade ago. The ratio for London, for example, had more than doubled to just under 7 since 1996. The report warned that less housing affordability is likely to lead to lower levels of home ownership, as is already evident in New Zealand.

Affordability is not the only problem. Interest rates and unemployment have risen from record lows with a commensurate impact on sentiment. Banks' lending criteria have tightened, especially following mid-2007 and the credit crisis—the requirement for larger deposits has choked the first-time buyer market. There are also concerns that the equity and rental returns on second homes, often bought speculatively in novel locations such as Croatia, Bulgaria, and Dubai, will disappoint, leading to a glut of properties and forced sellers. As mortgage arrears and repossessions rise, albeit from a low level, lenders further restrict the flow of cheap money. The fact that prices are now falling also serves to scare off buyers.

The boom in prices was spurred in good part by consumer borrowing, as discussed in the previous section. There is never a shortage of debate about the prospects for the housing market. Analysts said at the time of the market peak that property was overvalued by anything from 10 to 40 percent, and that further falls should be expected. There could be a major fall in house prices, or, less painfully, many years of no increases. Material nominal falls in house prices would be expected to bring on an economic recession as people grappled with negative equity, and the appetite for credit disappeared.

Others, notably some researchers from Oxford University, said that the higher price levels can easily be justified by taking into account housing supply, the changing age structure of the population, shifts in credit conditions and the level of nominal interest rates.[4] Based on their modeling, nominal house prices will only fall materially in the years ahead under "quite dismal" economic scenarios. A housing crash would bring widespread economic misery, while continuing high prices will consign the significant minority of the population that does not own property to a life of relative poverty.

John Kay, an academic and columnist, says that the level of house prices "depends not just on levels of income but on social mores and the distribution of wealth" and believes that modeling house prices requires a range of skills that few people have. That said, he has little difficulty with the concept of property hotspots. He explains the following:

> A house provides space and shelter and, in the American mid-west, these are the principal attributes of a house. There is more land there than anyone could build on and usually not much to choose between the prestige or convenience of different areas of the spacious cities. House prices are low, stable and tend to move in line with incomes.

In contrast, in New York, California, and London, most of the price of a property reflects the location rather than the accommodation; he points out that you cannot make more houses on East 69th Street or in Belgravia. With increased mobility, the high prices in the hotspots reflect "the self-defeating character of the search for the symbols of status and affluence."[5] House prices are, then, what economists call a positional good.

In the short term, policy makers have got their fingers crossed hoping for a "soft landing," typified by modest falls in prices, allowing the price-to-income ratios to return closer to the long-run average. Longer-run solutions to reduce the inequality brought about by the house price inflation are harder to find.

Some economists are discussing property taxes as a possible solution. In the States, increasing property taxes may be a slight deterrent, but they don't seem to have deterred anyone in recent years, so there's no reason to believe that an increase would solve things. The building of more homes would help, but land is limited in areas of high demand.

Meanwhile, expect house price crash stories to continue to appear in the papers. Prices will fall below the levels at the time of writing but when they have stopped falling we can be sure that life will still be tough for first-time buyers.

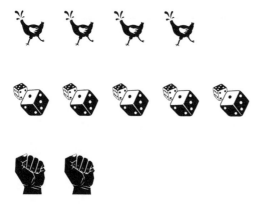

Immigrant Invasion

"Immigration issue reaching a critical point"
—*San Diego Union-Tribune*

Feelings about immigration run high—especially illegal immigration—with most newspapers being firmly entrenched on one side or the other. One U.S. editorial, titled "11 million reasons" (referring to the estimate of 11 million illegal immigrants in the U.S.) said, "The problem is not just getting worse, but way worse, and very fast." Another article, "False facts tar immigrants," said that some people in the immigration debate "regurgitate factoids ad nauseam, all with the purpose of blaming Mexicans for just about everything wrong with America."

Often, opinion runs along party lines. In 1998, president Bill Clinton voiced open support for immigration when he said that "America has constantly drawn strength and spirit from wave after wave of immigrants. . . . They have proved to be the most restless, the most adventurous, the most innovative, the most industrious of people."[1] Eight years later, Pat Buchanan drew much criticism for openly voicing concerns that "almost all immigrants today, legal and illegal, come from countries and cultures whose peoples have never before been assimilated into a First World nation."[2] It is unfair, however, to call Buchanan's extreme stance a typical Republican response, as the GOP is now making a conscious effort to move past intolerant statements of that type.

And the immigration issue is not just one for Americans to fret about. "Halt the tide of EU migrants," screamed the front page of one newspaper, demanding that Britain's borders be closed, while another warned that Britain's population would hit 70 million "unless [they] get a grip on immigration." One story quoted a "top military expert" warning that "migrant vandals will bring havoc to Britain," adding that mass population movements could lead Europe into a "Rome scenario," in other words the collapse of an empire. These may be headlines from tabloids, but the story of immigration and its impact on social structures and the economy is a favorite for all newspapers, whatever their hue.

Some of these stories are reporting sound research that merits careful examination. For example, the warning about Britain's population reaching 70 million came from Lord Turner, who studied population growth as part of his respected Pensions Commission report. He said that he was "amazed" by the "piece-meal discussion" on immigration and the "incoherence about the debate." He warned that high levels of unskilled workers entering the country might have a short-term benefit but would ultimately damage the economy, saying that "to deny that is nonsense—it just flies in the face of all economic theory."

Other stories lack substance, but there is no doubting that international migration has jumped up most countries' policy agenda in the last decade in response to the rapid growth in immigration flows. Globalization, budget flights, and more open borders, not least following the collapse of the Iron Curtain in eastern Europe and the broadening of the European Union, have increased travel opportunities and legitimate migration. Illegal immigration, via irregular or unconventional channels such as asylum-seeking, fake documentation, or overstaying legitimate trips as a student or a tourist, and humanitarian immigration, in response to civil and ethnic conflict, have also increased.

The numbers are large. Around 3 million long-term immigrants are recorded as entering developed countries legally every year. This is giving rise to strains as some ethnic communities are having difficulty integrating into the host society. Fear and prejudice freely breed in the absence of accurate and reliable data and a lack of coherent policy from governments—it is easy to fear the worst. In most countries there are serious gaps in available data with a too-heavy reliance on imperfect administrative information and poor-quality surveys. This affects policy makers, too.

There are definitional differences between countries—how long does someone have to stay to be counted as an immigrant?—making the compilation of international comparisons a tricky task. Surveys are notoriously weak as people are often reluctant to respond, let alone give the true reasons for traveling. But the size of the flows in the future is potentially enormous. One study, admittedly from a lobby group advocating lower population for the UK, raised the prospect of some 60 million people moving from desertified areas in sub-Saharan Africa toward northern Africa and Europe in the next twenty years.[3]

The lack of decent data, in this area at least, goes hand-in-hand with generally unclear government policy. Balancing an openness to international migration with a firmness in managing inflows is a difficult task for governments to achieve. Most do not even try—they dodge the issue. This in part reflects the sensitivities surrounding the topic—one incautious remark can lead rapidly to accusations of racism.

The OECD has been developing harmonized figures for "permanent-type" legal international migration that now cover nearly thirty countries and shed some light on the issue. Inflows were higher in most countries in the mid-2000s than at the start of the decade, notably due to a rise of one quarter over 2004 and 2005. The U.S. had by far the largest long-term inflow of foreign nationals in 2005, the latest year for which figures are available, of over 1 million. The UK followed with over 360,000 new immigrants, and Canada with 260,000. Germany, France, Australia, and Italy all received over 150,000 in the year. This represented an increase on the previous year for all these countries except Germany and France, which recorded small declines. When that year's inflows are presented as a proportion of the total population, it is the UK, Italy, and New Zealand that come top of the pile.[4]

These figures do not include short-term immigration, which is probably high in some countries that are prime "business trip" destinations such as the U.S. and the UK, and illegal immigration, which in common with any other covert activity is hard to measure with any degree of accuracy. Where estimates do exist, they put the illegal proportion of the population at between 1 and 4 percent in many developed countries.[5] The measure is highest for the U.S. and Greece, both of which are characterized by an extensive land border with a country of much lower per capita income—and nationals from the neighboring countries account for a large majority of the unauthorized immigrants. Such percentage figures might not seem very high, but it is striking that illegals represent a large proportion of the overseas-born population in many countries.

Whatever the tone of the newspapers—and the tone certainly runs the gamut—the flows are large enough to merit attention. Net unauthorized immigration to the U.S. is estimated to be around half a million persons per year, equivalent to the population of a decent-sized city and one half of the current annual levels of legal immigration, as measured by the issue of green cards. In

Italy, a regularization program—essentially an amnesty for illegal immigrants—in 2002 elicited 700,000 applications. If all of these entered in the years since the previous regularization, it would imply unauthorized entries of about 175,000 a year, higher than the recorded levels of legal long-term migration over the same period.

The situation is not easy for governments to manage as they do not have complete control over who comes into a country. This reflects both basic human rights (the right of residents to live with their families and the freedom to marry or adopt whom they wish) and a range of international agreements relating to refugees or the free movement of people. But governments do have control over the flow of some immigrants, and they choose to exercise this policy in very different ways. Australia and Canada, for example, select people on the basis of various characteristics such as language proficiency, work experience, age, and education. Their systems are transparent and often pointed to as models for other developed countries to follow.

Setting the quotas at the right level for the countries that pursue that route is another challenge—too many too rapidly presents difficulties finding work and integrating; too few and the potential benefits will not accrue to the host country, with labor shortages persisting and the best immigrants going elsewhere. Often where countries have set targets, there is little rationale for the chosen figure, it being the result of "political judgment" with the consequence that they fail to attract wide support. Temporary migration, based on a permit system, might be appealing to a skeptical public and might be acceptable for some categories of low-skilled workers, but such newcomers are likely to be less adaptable and integrate more slowly. Ongoing, regular labor needs are unlikely to be met most satisfactorily by recycling temporary workers.

A major focus of controversy in Europe since 2004 has been the impact of the accession of ten countries to the EU in May of that year, taking the number of members to 25. Membership was increased again at the beginning of 2007 with the accession of

Bulgaria and Romania. With membership came the free movement of people and an extensive debate up to accession surrounding the likely numbers of eastern Europeans—the population of the ten countries was about 73 million—who would choose to move west.

Initially, only Ireland, Sweden, and the UK offered unrestricted access to their labor markets for the citizens of the new EU members. Perhaps because of the restrictions in other countries, the inflow of immigrants to them, and the UK in particular, was much larger than expected. More than 170,000 people from eastern Europe applied to work in Britain in the first 11 months, and others came without permits, much higher than the official government estimates, made before accession, of between 5,000 and 13,000 in a year. The gap between the forecast and outturn was very damaging to the government's reputation for policy competence, and the size of the inflow led to arguments about funding for public housing and schooling. Two years after accession, the government admitted that about 600,000 people from the new EU countries had come to work in Britain. Over half of those who came were from Poland. One government publication said that 2005 "saw the largest ever entry of foreign workers to the UK, totaling some 400,000."[6] The distribution of the new workers within the country was very uneven, with nearly half going to London. The emigration was so large that it has caused concerns in the countries of origin.

Much of the furor surrounding immigration, which can be a significant and beneficial activity, reflects the media and public attitudes to a relatively small number of system abuses by asylum seekers, people traffickers or others engaged in illegal activity—and might well have a racist undertone. In the early 1990s, the possibilities for migration to developed countries were very limited, but the successful flight of refugees quickly focused attention on asylum seeking as a means of entry for those seeking a better life.

The vast majority of asylum seekers—up to 90 percent in many periods in many countries—fail to have their claim for asylum recognized, reducing the legitimacy of this route. The number seeking asylum in developed countries has almost halved compared to 2000, but still amounted to 300,000 in 2005, with over 25,000 asylum seekers in each of the U.S., UK, France, and Germany. The very high and rising levels in the 1990s put national processing systems and public sympathy under considerable strain. Even so, refugees and other persons admitted for humanitarian reasons, and their accompanying families, currently account for no more than 10 percent of long-term immigration in developed countries.

Abuses of the system attract large headlines and strong public criticism. In 2006, a group of 150 HIV-positive women, mainly from South Africa, Eritrea, Uganda, and Zimbabwe, refused to leave Canada, seeking asylum, having attended an AIDS conference in Toronto. Back in 2000, an Afghan airliner on an internal flight was hijacked and diverted to London's Stansted airport, at which point sixty of the 150 passengers claimed asylum. People who do not have the entitlement to stay in a country might opt to enter into a bogus marriage. It is estimated that there were 5,200 "sham" marriages in the U.S. in 2006, contributing to a multi-million dollar industry in this country.

Half the developed countries now have foreign-born populations representing at least 10 percent of their total populations. Adding offspring can take the percentage with a recent immigrant background to 40 percent or more in some countries. In practice, figures are likely to be larger since some legal immigrants and most illegal immigrants will not be counted in the official population figures. Illegal immigrants often cannot join their municipal population register, frequently the source of the numbers, as they do not have a valid residence permit, and often they will choose not to complete a census form, concealing their presence rather than risk being found. Foreign-born residents make up over 20 percent of the officially counted population in Luxembourg, Switzerland, and Australia, about 13 percent in Germany and the U.S., and a little less in the UK, France, and the Netherlands.

Concern about the lingering uncertainty of the numbers is compounded in people's minds by the uncertainty over the economic contribution of immigrants. The newspapers do not help to clarify the situation. On the one hand, we are told "80 percent of migrants are a net drain on the economy," and on the other, "Spanish study points to benefits of immigration," arguing that national per capita outputs might have fallen in the last decade instead of increasing at a healthy rate had it not been for influxes of immigrants. Whatever the truth, the rapid change in some parts of some cities due to the high levels of recent immigration has been unsettling to many long-term residents.

Losing Control of Your Vehicle

"Investigators: Nascar plane crash was avoidable"
—*The New York Times*

Illicit cargoes, human or otherwise, produce their share of scare stories, but the modes of transport they choose are not regarded by the media in quite the same way as many other hazards. Although some people are genuinely terrified of particular forms of transport, especially flying, their fears are not translated into headlines anticipating disaster as they are for diseases or violent crime. Why is this?

The obvious answer seems to be that transport disasters actually happen often enough to satisfy the media's demand for excitement—with speculative alerts before the event replaced by righteous fulmination after it. Each nation has its own roll call of disasters recalled in locations of rail crashes, names of ships that have sunk, or the flight numbers of planes lost.

A contradictory explanation for the absence of scare stories is that transport is too much an essential part of daily life for us to entertain them. The evidence of our continued survival of the daily car or train journey gives them the lie. Travel must be safe.

Both of these perceptions are false. Moving around is intrinsically dangerous. Transport accidents are the leading cause of accidental deaths after falls. In the U.S., 4 million traffic accidents a year result in around 42,000 fatalities, which, in addition to the senseless loss of life, incur direct economic costs of $230.6 billion a year.[1] Yet when measured by the distances covered and other gains to the traveler, travel is considered pretty safe, and it's getting safer all the time. Its high benefit-to-cost ratio makes us prepared to accept the real dangers of travel. This pill is sweetened by the fact that transport often lacks the essential "dread" factor that arises in the case of risks that are "globally catastrophic" or where "little preventive control" is possible.[2] A transport disaster, by contrast, mainly affects those taking the transport and does have the possibility of preventive control exercised by a human driver.

Nevertheless, there are reasons why we might be alarmed. We travel ever greater distances, whether commuting to work, taking cheap flights on vacation, or making longer round trips to grander but more remote supermarkets and other facilities. It is natural to ask how safe we are. They may not inspire dread, but transport accidents do have some unpleasant characteristics that mark them out from other risks. They occur across the board to all age groups and social classes. It is therefore not easy for one community to reassure itself of its own safety because the risk is greater for another, as may be the case with illness or violence. Transport accidents are also responsible for more years of lost life than many other accidents—people tend to fall off ladders in their dotage but off bikes in their prime.

The Bureau of Transportation Statistics collates deaths in the U.S. per 100 million miles traveled for many modes of transport. In recent years, crashes of U.S.-registered airlines have resulted in fewer fatalities each year, with a low of 1.37 for every 100,000 departures in 2005. General aviation, by contrast, resulted in 2.43 deaths for every 100,000 departures, meaning that non-commercial aviation, which is clearly bringing the average up, is much more dangerous. Large trucks (big rigs) caused only 0.36 deaths per 100 million miles traveled, as compared with 1.14 for both passenger cars and pick-up trucks.[3] BTS neglected to report statistics for a lot of other modes of transportions, but statistics from the UK's Department for Transport (converted into miles) can help us fill in the picture a bit: travel by water 0.05, by rail 0.06, cycling 6, walking 8.[4]

Except perhaps to a statistician, these figures at first seem impossibly low. One hundred million miles is an awfully long way. With 1.14 deaths per 100 million miles for car travel, it seems you can expect to drive nearly 90 million miles, on average, before you meet your maker. Is car travel really this safe? How come there is carnage on the roads? The answer is simply that the roads are so busy. It is all car users taken together who rack up this number of miles between fatalities. There are over 210 million cars in the U.S., which travel an average of 11,000 miles each per year. Based on these figures, we can calculate the average total annual death toll suffered by car occupants as 1.14 per hundred million × 210 million × 11,000. This comes to 26,334. The official figure of around 42,000 deaths annually covers all *road* accidents in the U.S. including those involving other forms of transport as well as pedestrians.

Different forms of transport are often in competition. We face a choice of flying or taking the train on vacation. We drive, cycle, or walk to the store. This competition extends to the way that respective transport industries choose to present these already opaque statistics. Where journeys are long, it is advantageous to quote the accident rate per distance traveled, as above. Airlines come out safest by this measure. Where journeys are short, it is better to quote the accident rate per journey, which is how the UK authorities report their air travel statistics. But the FAA, realizing that commercial airlines safely carry so many passengers, reports the statistics on a "fatalities per passengers on board" basis.

Motorcycles come out as the most dangerous means of transport of those surveyed, with 40 fatalities per hundred million miles—almost forty times more dangerous than cars. The only regular form of transport popularly believed to be more dangerous than the motorcycle is the helicopter; U.S. civil helicopter travel has a fatality rate of 6,200 per billion journeys.[5] Even worse, there has been a recent spike in helicopter fatalities: The *Wall Street Journal* reported in 2008 that helicopter travel is now 300 times more dangerous than other forms of air travel, and that at least 35 lives had been claimed by emergency medical transport missions since 2007.[6]

These data may not be strictly intercomparable since the statistics tend to be recorded in slightly different ways for each form of transport. However, they do permit some broad conclusions to be drawn. As the result for motorcycles highlights, we face the greatest risks when we ourselves are at the controls. Per mile traveled, motorcycle, foot, bicycle and car are the most dangerous forms of transport, with rail and air the safest. According to the risk psychologist Paul Slovic: "The public seems willing to accept voluntary risks roughly 1000 times greater than involuntary risks at a given level of benefit."[7] If this is anything like truth, then car travel, which is only around nine times as dangerous as going by train, is a freedom that we are unlikely to relinquish. This may help to explain why, despite a high level of preventable fatalities, initiatives to improve car and road safety progress only slowly.

The situation is very different in UK rail travel where a rapid sequence of tragic accidents—Clapham, Southall, Ladbroke Grove, Hatfield, Potters Bar—has focused a harsh media spotlight on "failings" in the industry. These events attained a visibility arguably greater than that warranted by the loss of life incurred, partly because they provided a focus for broader public concerns about the effects of the recent privatization of the railways. Even including these disasters, the record of UK railways before and after privatization is one of continuing safety improvement. Up until 2006, the Health and Safety Executive monitored the number of "significant incidents" on the railways, including collisions and derailments affecting passenger services, but excluding accidents involving level crossings. These have fallen from around one significant incident per million journey miles in 1975 to 0.2 per million miles today.[8]

The statistics reflect the unequal pressures on private and public transport. All the accident rates surveyed by BTS have improved compared to those for the previous decade, except for motorcycling which has become marginally more dangerous.[9]

Car travel in general has become consistently safer in developed nations. A rare opportunity to make a long-term comparison came when the German Federal Statistics Office decided to mark its centenary of keeping track of road accidents. In 1907, there were 4,864 recorded traffic accidents in Germany, in which 145 people were killed and 2,419 injured, out of a car population of just 27,026. In 2005, 5,362 people were killed on German roads, but there were 56.3 million cars. The comparison led the Reuters news agency to draw the conclusion: "German drivers 56 times safer now." In Europe, deaths per kilometer driven have been declining steadily by about 4 percent a year as the combined consequence of increasing car safety, safer roads, and tougher driving tests.

Turning to public transport in the U.S., the fatality rate for buses has decreased slightly: In 2003, bus travel was responsible for 0.09 percent of transportation fatalities, down from 0.1 percent the year before. Water transport too is down ever so slightly: from 0.28 percent in 2002 to 0.27 percent in 2003. Over the years, the safety of personal transportation by car has made the greatest progress, but is nonetheless one of the most dangerous methods of transport.

The perception of who is in control is important in determining how we respond to transport dangers. A lone car driver takes an entirely voluntary risk when he or she gets behind the wheel. That person's passengers take an involuntary risk if they share the journey. However, this involuntariness is mitigated by the fact that they can see the driver and can take a view as to how reliable he or she is. Passengers in a train or a plane face an additional involuntary degree of separation because they must put their trust in an unseen driver. In reality, this driver may have little operational control and may only be there at all as a figurehead to reassure the public, so that passengers are in practice putting their trust in a *system*.

In September 2006, a German magnetic levitation train crashed while on a high-speed test run near the Dutch border. Most of the 29 passengers on the experimental journey were killed, along with two employees of the company developing the train who were on the ground. "At least 21 die as driverless train crashes into maintenance truck," reported the *Guardian*. The key word for the media here is "driverless"—it was repeated in the first sentence of the story and in a text highlight. (The journalists who went to work that day on certain stretches of the London Underground were presumably unaware that they too were riding in driverless trains.)

The implication is that a driver would have seen the truck and stopped the train—which is almost certainly not the case at high speeds. Such perceptions will need to change if we are to gain an accurate impression of travel hazards in the future. Developments in transport technology already make travel generally safer. Would-be disaster headlines now betray small miracles. MSNBC told of a "Miracle on the Hudson" after a pilot managed to land a wayward craft safely in the Hudson River just west of midtown Manhattan in 2009. "Jets fly on despite engine failures" was the *Sunday Telegraph* headline accompanying "news" that more than 20 British passenger planes had not crashed on long-haul flights over the previous five years when they had suffered problems forcing pilots to shut down an engine.

Airlines are now keen to make flying cheaper by using planes without pilots. Flight control technology makes this feasible, but it is judged that the public is not yet ready for this innovation. On the roads, the decline in car fatalities could become much sharper as cars too are equipped with computer devices to monitor and control driving. A lot of car enthusiasts will undoubtedly complain that such measures cramp their style, and there will equally undoubtedly be aggrieved tales of unfortunate people who suffer accidents nevertheless and who now hold the new technology to blame for them. But overall, accidents and lives will be saved if we can bear to let go of the controls.

Death by Phone

"Why cell phones and driving don't mix"
—*ABC News*

Death on the road, perhaps in dramatic circumstances and often involving children or young people, with a phone as a contributory factor gets expansive coverage in the media. "Was death crash girl texting at the time?" is an example of the sort of tragic story that makes heart-wrenching reading. In one case, a truck driver drove into the back of a stationary car while punching the keypad of his new phone, killing a young woman. He admitted to dangerous driving. In another, it was the person suspected of using the phone who was killed, leaving the relatives fighting for the person's reputation. The common theme in the stories is that people's lives are ended as a result of a moment of stupidity or carelessness.

What are the facts behind the emotion? Cell phones have been around for a long time, but it is only during the last decade that their use has become widespread among the population. Comprehensive network coverage means that they are increasingly used by drivers of vehicles. In an increasing number of countries, it is now illegal to drive while using a mobile phone. So far in the U.S., nine states and the District of Columbia have outlawed it.

But just how dangerous is it to drive while using a cell phone? While it must be more dangerous to drive using a cell phone than to drive not using a cell phone, how does this particular in-car diversion compare with others? Once again we are faced with imperfect data. Information has not as a rule been collected at accident sites to assess whether cell phone usage might have been a contributory factor—robust data would in any case be hard to collect. Breathalyzer tests are now conducted as a matter of course in many countries when an accident has occurred, improving our knowledge enormously about the role that alcohol plays in such events. But as yet, information on other stimulants, such as drugs, or diversions, such as cell phone usage, is not in general collected at the accident scene. The number of accidents where cell phones are known and declared to have been a contributory factor are very small but in all probability are just the tip of the iceberg.[1]

Accordingly, we are left relying on specially conducted research. One early and much-quoted example in the *New England Journal of Medicine* in 1997 looked at the cell phone call records of nearly 700 drivers who had been involved in vehicle collisions resulting in substantial property damage.[2] It concluded that the risk of collision when using a cell phone was at least 3 to 6½ times higher than when one was not being used. It concluded that the relative risk was similar for drivers who differed in personal characteristics such as age and driving experience, and was the same during the day as the night, in winter and in summer.

"Mobiles worse than drink driving" was how one piece of research conducted by Britain's Transport Research Laboratory was reported, which suggested "talking on a mobile phone while driving is more dangerous than being over the legal alcohol limit."[3] This research involved a panel of 20 volunteers using a driving simulator to measure reaction times and stopping times while using a cell phone and under the influence of alcohol. The research said reaction times when talking on a cell were, on average, 30 percent slower than when just over the legal alcohol limit, and nearly 50 percent slower than when driving sober. Drivers on the phone were also less able to maintain a constant speed and found it more difficult to keep a safe distance from the car in front. There is no reason not to believe the results of this test, although without more details and bearing in mind the small number of subjects— and the knowledge that the study was sponsored by an insurance company—the detail of the results should perhaps be taken with a grain of salt.

Some of the other research might be a bit less robust and therefore more open to misinterpretation. In the UK, where cell phone use while driving has already been banned throughout the whole country, one poll of over 1,000 motorists was reported under the headline "10 million ignore the law on phone calls at the wheel."[4] Only 58 percent of those polled said that they fully observed the law and did not use their phone when driving—grossing up the remaining 42 percent leads to the figure of 10 million. The article went on to say that the figures "suggest that driving while using a handheld phone is 20 times more prolific than recent Department of Transport research claimed." This phrase alerts the reader to a yawning gap in the message coming from two different data sources, adding to skepticism and concern, but unfortunately in this case probably erroneously. The official figures come from a survey, conducted in April 2003, suggesting that 2.4 percent of drivers, roughly 1 in 40, are using a cell phone *at any moment in time.*[5] The two figures are very different, but both are plausible as they are measuring different things.

One of the more interesting conclusions from both the *New England Journal of Medicine* and the Transport Research Laboratory studies is that hands-free kits were almost as dangerous as handheld phones. The former study concluded that hands-free units "offered no safety advantage over handheld units." The latter study said that under normal circumstances the braking distance for a car traveling at 70 mph, the speed limit on Britain's motorways, is 31 meters (101 feet). This increased to 35 meters (115 feet) with alcohol, 39 meters (128 feet) with a hands-free phone and 45 meters (148 feet) with a handheld phone. It is interesting to note that, despite this evidence, legislation and public opinion in virtually all countries ignore the risk from hands-free phones while punishing those found to be driving using a handheld phone or over the limit for alcohol.

So if it is conclusive that the use of cell phones while driving is more dangerous than not using cell phones, why do governments—and pressure groups for that matter—shy away from the prospect of a proper risk assessment of other in-car activities? Perhaps it is because any such assessment could well conclude that there are many more practices and conditions—and perhaps even people—that should be banned. Almost any form of in-car diversion, physical or mental, such as talking, controlling children, map-reading, loud music, smoking, or eating, ought to be studied and, if appropriate, be banned.

There are also other less obvious risks while driving. One such example, resulting from research conducted in the Czech capital, Prague, suggested that the significant minority of the population with latent toxoplasmosis—cysts in neural and muscular tissues—had an odds ratio of between two and three times higher risk of an accident. And often accidents are not the fault of the vehicle driver. Many of the road accident deaths are of pedestrians, some of whom may have been talking on a cell phone while crossing the road. Would it be sensible to introduce laws to dissuade people from speaking on the phone while crossing the road?

The research on driving while using cell phones is more advanced in America than elsewhere. One piece of research estimated that each day, cell phones in the U.S. accounted for nearly 1,000 reported collisions, 1,700 total collisions, 317 persons with injuries, 2 deaths, and 99 lost years of life expectancy.[6] The research concluded that regulations controlling the activity "may be justifiable because the benefits and harms do not always involve the individual who has a cellular telephone," but also noted that other regulations, such as limiting the access of teenage males to cars, could be more cost-saving to society.

The risk of being killed as the result of a road accident caused by the driver using a cell phone is not large. But even if the number of these types of accidents stabilizes or declines, it seems likely that the keen human interest element in such stories means that they will still feature strongly in the media.

There are two other factors that are likely to see such stories prosper. The first is the refusal of a minority of the population to accept the new laws. A *National Post* story headlined "Selfish cell-phone users have Toronto police fuming" quoted a number of senior members of the community explaining in forthright manner why they would not support a ban. One even suggested that he would set up a business selling tinted windows so that those using their phones while on the move could not be spotted by the police.

The second is the advent of new technology and the application of existing technology for use in the car, which could lead to a whole new round of scare stories. A report in a Seattle-based newspaper suggested that a significant minority of local residents already had Internet access in their cars and used it for e-mail and web surfing, sometimes while driving. The rising use of DVD players and gaming machines in cars is also causing concern as some drivers are apparently dismantling the systems that prevent them from playing while the car is moving and installing them in a place, such as the dashboard, where the driver can see the screen. The article reported the case of a man in New York who was stopped for watching pornography on a screen in his passenger-side sun visor. It wasn't the simultaneous driving and viewing that the state objected to, but the public display of sexual material!

5. The Workplace

Money makes the world go around, and when we perceive a threat to our livelihoods we tend to panic. Globalization is a relatively new cloud on the horizon, but discrimination of one sort or another—we discuss that between men and women here—has been around for much longer. Too much work can lead to stress—and of course work itself can be dangerous.

No Work or Low Pay

"Offshore jobs in technology: opportunity or a threat?"
—*The New York Times*

"Tech workers' losing fight to match overseas wages," was the headline in 2003 when U.S. workers began losing technology jobs to outsourcing. "It's a total deflation period," a Forrestor Group research director said of the state of affairs for techology workers. And manufacturers are experiencing similar problems: One worker faced with downsizing at a Detroit engine plant said, "It's a helpless kind of feeling." His union leader said, "The Chinese are coming."

One of the unwritten rules of capitalism is that the workers in companies with rising profits are rewarded with higher pay. But this tacit arrangement has taken a new twist as multinationals across the industrialized world enjoy very high profit growth while the wages of many workers in their home markets in the West, and especially those on low earnings, are failing to keep up with inflation. The hit has been felt in the last couple of years, when taxes, interest rates, and energy prices have been rising, straining household budgets. Income growth is being secured as a result of this corporate success but by the peasants-turned-workers in the developing world. While the working classes in the West might struggle, those at the top have been doing fine: Owners and directors have enjoyed large pay increases and benefited from stock options and performance-related bonuses, while the financial community and investors gained from the buoyant mergers and acquisitions activity and share price rises.

The shrinking range of especially manufacturing jobs and stagnation of blue-collar wages in the West has arisen, as ever in the world of economics, due to a complex cocktail of events, but globalization has played a large part. In the good old days, western workers enjoyed a near monopoly access to western investment: American companies generally invested in America and British companies in Britain and they employed local labor because there was nowhere else to go. Most of the products were sold in the country of origin, but some were exported. Globalization, however, has changed that as work has been exported from developed countries. As a result, "Distrust of globalization has probably never been higher in the past 60 years," according to one Canadian columnist.[1]

It has been estimated that, as globalization has taken hold over the last two decades, the number of workers in what might be described as the market economy has doubled from roughly 1.5 billion to nearly 3 billion, mainly as a result of the opening up of China and India, but also of the fall of the Iron Curtain in the early 1990s. The basic appliance of the rules of supply and demand would suggest that such a large increase in the supply of labor in a relatively short time, with output growing more slowly, will lead to job losses or wage cuts where wages are highest. It is natural that the bargaining power of labor declines in such circumstances, a position exacerbated by the fall in the strength of trade unions.

One American columnist noted the following:

> Globalization isn't working out as promised ... Americans were assured that new trade accords and China's member-ship of the World Trade Organization would mean better living standards for American workers. That's because China

and other countries supposedly would buy American exports. In reality, American manufacturing jobs have been decimated, and the U.S. is running an unsustainable and destabilizing trade deficit, especially with China.[2]

The phenomenon of job exporting started in manual, blue-collar jobs but is now spreading to many professions. In the 1980s and early 1990s the low earners fell further behind the middle class while the rich pulled ahead. Now the poor are keeping pace with the middle, but the rich are pulling away. The share of total income captured by the top hundredth in the U.S. doubled from 8 percent in 1980 to 16 percent by the middle of this decade.

It is no wonder that most of the nervousness about globalization is directed at China. It accounts for one in five of the world's population and is more than four times the size of the U.S. and three times the population of the European Union. Few in the West have been to China; even fewer can communicate in Mandarin. The mystery is enhanced because of the country's traditional policy of secrecy, which began to unwind only twenty-five years ago with a process of economic liberalization.

On all accounts—although the quality of official statistics is notoriously very low—China has made considerable progress during that time: The economy has increased twelvefold, and trade with the rest of the world by a factor of thirty. In the last decade China's GDP has overtaken that of Canada, Italy, and the UK, and is on course to overtake that of Germany soon, making it the world's third-largest economy after the U.S. and Japan. These figures, however, grossly understate the real size of the Chinese economy once adjustment is made for the country's relatively low cost of living. Such adjustments suggest the Chinese economy is perhaps four times larger than the crude figures—an estimate supported by China's capacity for electricity generation, estimated at around five times that of the UK.

China, and India too, are rapidly moving up the technological ladder. China is already graduating over 1 million scientists and engineers each year, and it is estimated that, by the end of the decade, China will graduate more PhDs in science and engineering than the U.S.[3] The plummeting cost of telephone calls means that all sorts of service-sector jobs, not just call centers, can also be exported. And although the negative consequences of the growth in China and India are currently focused on the very expensive labor markets of developed western countries, they have also been affecting workers in Latin America, South Africa, and the more developed areas of Asia too.

The impact of globalization is far more pernicious than the shifts in trade or industrial structure we saw a generation ago. Then, one industry such as coal mining would be in decline, or one firm would close or relocate, but now the impact on the West is much wider than this. As a constant flow of manufacturing workers in the U.S. lose their jobs as Chinese imports rise, they will seek employment in other sectors, such as retailing or construction, depressing wages in those sectors too. Some blue-collar workers find this transition and the required retraining hard, but accountants, sales people and drivers, for example, will find a switch to other sectors more manageable.

The downward impact on the income of many households would have become more apparent in the last decade had it not been for rising house prices (making many people feel richer), rapid increases in consumer borrowing and the increase in the economically active proportion of the working-age population. It is also true that the industrialization of China and other countries has prompted dramatic deflation in the prices of many goods. In the U.S., for example, the overall consumer prince index fell sharply toward the end of 2008, with the biggest change being in the average price of IT equipment. These offsetting trends could not continue for ever, and now they are diminishing, the disadvantages of globalization and technological advance will become more apparent to more people.

It will then become clearer that globalization is about massive waves of income redistribution: from workers to consumers, as they can shop around ever more widely for cheaper goods; from expensive labor to cheap labor, as employment expands rapidly in developing countries; and from energy users toward energy producers, as the demand for energy soars in developing countries.

The industrial structure of countries is constantly changing. The decline of traditional industries such as coal mining, iron and steel, and textiles and clothing was, of course, firmly in place for decades in the U.S. and elsewhere before it became fashionable to talk about China. Now unfashionable state intervention often masked or delayed the decline, but the trends were clear. In the U.S., manufacturing employment was at 17 million in 1997 and is now just over 14 million.

Not all industry has declined in the West. The share of finance and business services, for example, saw a fivefold increase in the U.S. between 1990 and 2005. Even though heavy industry has been decimated in the West, other sectors, such as publishing, pharmaceuticals, and motor vehicles, are expanding. The statistics can also deceive—when a manufacturing company outsources its IT, catering, or cleaning, the manufacturing sector appears to decline while business services increase, yet the same quantity of goods is being produced by the manufacturer.

Workers in developed countries do not panic about globalization on a week-to-week basis, but the medium-term consequences could be considerable. Alan Blinder, a former U.S. Federal Reserve vice-chairman, said about off-shoring in a congressional testimony, "We have seen only the tip of a very big iceberg." Adding, "Tens of millions of additional American workers will start to experience an element of job insecurity and downward pressure on real wages that has hitherto been reserved for manufacturing workers."[4] He predicts that the key labor market divide going forward will not be between high-skilled and low-skilled workers, but between services that can be delivered electronically from off-shore and those that cannot be. He estimated that between 22 and 29 percent of all current American jobs might potentially be offshorable. Perhaps people in the West should become waiters or brain surgeons rather than typists or researchers.

The problem is that, beyond words, there is relatively little that governments can do. Social benefits can be increased at the margin for some of the victims, but sharply increased public expenditure is not a long-term solution for aging societies with rising unemployment. Taxes could be increased for shareholders, company owners, and the senior executives who are profiting from globalization, but globalization has simply made entrepreneurs and their capital more mobile. They can be expected to move to the areas of lower taxation if one area were to raise their taxes.

We might expect the calls for trade tariffs and protectionism to intensify and become more widespread. And expect a range of social concerns to be raised—should we be trading with countries that allow child labor and have inadequate workplace safety regulations? While such social differences between the developed and developing world remain, labor costs in the West are always going to be higher than elsewhere.

Underpaid Women

"The gender pay gap: still alive at the top, too"
—*BusinessWeek*

The evidence appears to be compelling. A headline such as "Scant progress on closing gap in women's pay" vividly sets the tone, and one American lobby group has claimed that "the typical female college graduate in 1984, who is now in her mid-forties, has lost a total of $440,743 in the years between 1984 and 2004" due to the gender pay gap.[1] But are working women *unfairly* paid less than men?

Despite the strident nature of these headlines, the figures suggesting that women earn significantly less than men are not all they might seem at first glance. There are a range of statistical issues to consider. First, there is the difference between median earnings (half the people earn more than this and half earn less) and mean earnings, with the gap based on median earnings invariably being smaller than that based on mean earnings. The median removes the impact of the very high earners, who tend to be disproportionately male, but it is preferred by most analysts as it gives a better indication of the situation for a typical person—lobby groups use the average as it gives a larger figure.

Second, women's total weekly earnings, measured in dollars, will be below the level achieved by their male counterparts as full-time men tend to work longer hours than full-time women. Third, the impact of part-time workers on the pay gap can be significant as part-time women are disproportionately represented in jobs where the pay is around two-thirds that of part-time men for every hour worked.

The gender pay gap also widens as any given cohort of men and women grows older. For school leavers the gap is negligible but it rises steadily up to the age of about forty. This is partly because as women get older they are more likely to have spent time out of the labor market caring for children or elderly dependants. There is some evidence that the gap between male and female earnings declines again during people's forties and fifties due probably to the pay of very high-earning men slipping back.[2]

Despite the complexity of the measurement issues, the gender pay gap figures are often presented as if the gap is due purely to "discrimination" against women. According to the National Organization for Women:

For full-time, year-round workers, women are paid on average only about 77 percent of what men are paid; for women of color, the gap is even wider. These wage gaps stubbornly remain despite the passage of the Equal Pay Act more than 40 years ago, and a variety of legislation prohibiting employment discrimination.[3]

On closer analysis, it becomes clear that only a small percentage of this gap arises out of what might reasonably be called discrimination.

Pay discrimination occurs when a man and a woman, who have similar qualifications and are doing similar jobs, are paid different rates. Because the pay gap as measured by statisticians is an aggregate concept, not matching person against person, there are lots of reasons other than what most people would see as discrimination that could explain, and justify, its existence. Most notable among these explanations are differences in the pattern of male and female employment (in other words, they do different sorts of jobs), previous employment histories, childcare responsibilities, and levels of qualifications.

It remains the case that many women during their twenties and thirties will take some time out from work to have children, and recent legal cases have supported the common-sense argument that a 40-year-old man with a full employment record with the corresponding experience can expect, other things being equal, to be paid more than his female counterpart who has taken a long period out to have children.

Each generation of women has done better on pay relative to men than the previous generation, but it is true that the pace of improvement has slowed in the last decade or more. One possible explanation for the stabilization of the pay gap could be that discrimination, as defined in the wider world away from lobby groups, no longer exists. It wasn't until 1850 that married women began to have full ownership of their own possessions and wages. And before the 1963 Equal Pay Act, it was routine for collective agreements to have women's pay rates below those of men. Probably, the gap now between men's and women's pay largely, if not totally, reflects factors other than discrimination.

In the absence of any figures from governments that highlight pure discrimination, a number of academics and lobby groups have conducted research attempting to analyze the relative importance of the various factors contributing to the causes of the gender pay gap. Even so, it is difficult to come up with a definitive breakdown. The Equal Opportunities Commission in Britain published research that identified four main causes of the pay gap: discrimination, education, labor-market rigidities, and working patterns. They believe that discrimination, which includes what they call "systematic disadvantage," in other words, the work women choose to do being paid less than the work men choose to do, accounts for just over one-third of the pay gap. Another third was due to different lifetime working patterns, in other words, the tendency to work part-time and to look after children, and about one-tenth was due to women spending less time in education.[4]

Identifying and measuring these reasons is very difficult, and such studies are controversial. One American (female) researcher concluded that "There is no proof that discrimination is the cause of the remaining pay gap. It is possible that the average man, brought up to view himself the main breadwinner, is more committed to his job than the average woman."[5]

One piece of research suggested that a significant part of the difference might be accounted for by the fact that women "are too scared to ask for a pay [raise]."[6] The survey said that two-thirds of women had never put in a request for more money even though 80 percent believe they are underpaid. Men, in contrast, were generally much pushier when asking for money. This specific example would support the view that there are systematic differences in personality—essentially that many women are less competitive than many men, tend to be less self-confident, less motivated by money and less effective in negotiation—which contribute to the choices made in life that can lead to the existence of the pay gap. One survey suggested that four in ten women aimed to be financially dependent on their partner, relying on his income, savings, and pensions to be secure in the future.[7]

Some of the differences between the sexes are visible long before the young people have entered the workplace. Women have now reached parity with men in terms of obtaining college degrees—and by some counts have surpassed men—but as a study by the American Association of University Women (AAUW) found, women still tend to obtain degrees that quality them for "pink collar" jobs like teaching and nursing. In college, men consistently outnumber women in computer science and engineering courses, while women outnumber men in education, languages, and subjects allied to medicine (excluding medicine itself and dentistry). Mary Ellen Smyth, president of the AAUW's Education Foundation, says "it's not that women are hitting a glass ceiling in the high-tech sector. It's that they don't have the keys to open the door."[8]

There are very few occupations that have roughly equal numbers of men and women. Men dominate in computing, the police, and engineering, while women dominate in office support staff, primary-level teachers, care assistants, and hairdressers. It is impossible to know whether the pay in the jobs dominated by women is lower on average because women dominate.

The pay gap in America has, however, been narrowing, but arguably not for the reasons anticipated or hoped for. The blue-collar and manufacturing jobs where men dominate have seen considerable downward wage pressure for more than a decade, while women, who are more prevalent in service-sector jobs such as healthcare, have seen wage rises, albeit from lower levels. One article, "Rise of the women who earn more than their men,"[9] told us that almost 40 percent of women who work full-time are paid more than their husband or live-in boyfriend, hinting at progress being made by women in plugging the remaining gap.

But those who lobby to close the gender pay gap argue that new policies, such as compulsory pay audits for all organizations and legislation to tackle the culture of long working hours, which tends to discriminate against women, are required.[10] They argue that it is important "to challenge the myth that gender equality has now been won and . . . [believe] that continuing gaps are the result not of free choices made by women and men but of an unequal society."[11]

If, though, the blatant forms of discrimination are no longer contributing much to the gender pay gap, sensible policy initiatives to close it are hard to find. Britain's Women and Work Commission published its "Shaping a fairer future" report in 2006 and concluded that women are not making full use of their skills and that the primary cause is the culture of schools and workplaces.[12] Policies such as reducing gender stereotyping in schools and encouraging women to study male-dominated subjects, and making more senior jobs available to those who want to work part-time, are worthy initiatives but hard to implement. Their success is hard to measure.

Perhaps the biggest culture change required to close the gap is not in the attitudes and aspirations of women but in those of men. Perhaps a society with more male home-makers would be more appealing than one with more female truck drivers. Meanwhile the stage looks set to remain clear for the chauvinist comedian who believes in equal pay. He must be a believer in fair pay, he says, because he lets his female staff work longer hours so that they can earn the same as his male employees!

It's All Too Much

"Stress triggers heart-damaging behavior"
—*Health*

"It is a profound privilege to die from stress-related diseases," says a professor from Stanford University. The point he makes, of course, is that in developed countries we have never had it so good, and that worrying about stress is itself a sign of how charmed our lives are. As a society we have wealth, job choice, and travel opportunities unimaginable only a generation ago, and in our free time we can gamble, drink, surf the Internet, and watch television on super-sized plasma screens to our heart's content. We have legal safeguards against many of society's ills, and the hard toil and infectious diseases that filled the Victorian graveyards with youthful corpses have all but gone. And yet it seems we are as miserable as sin and bogged down with stress.

Stress is, apparently, a serious problem, at an all-time high, killing us or costing us a fortune and putting companies out of business. There seems to be no corner of society, no group of people, that avoids being afflicted. In a survey from the American Psychological Association, 73 percent of subjects reported feeling stress-related psychological symptoms such as irritability, anger, nervousness, and torpor in the previous month. Seventy-three percent believe that stress can make you physically ill, and 79 percent felt that stress is a fact of life. Forty-five percent of those surveyed felt that stress was causing problems in their relationships.[1]

The above survey and numerous others have confirmed that work is the leading cause of stress. Nineteen percent of responders in the 2000 Integra Survey said they had in the past quit a job due to the stress, and 12 percent admitted that stress had caused them to call in sick.[2]

There have been many other surveys on the topic of stress, giving the impression that almost every trade union or professional association has consulted its members on the topic. In addition, a good number of tables purporting to rank jobs according to their stress level have been produced and are a favorite for the media.

Business leaders in Sweden should spare a thought for their counterparts in Taiwan. Apparently only one-quarter of the former believe that stress levels are increasing, compared to 90 percent of those in Taiwan. Stress levels were broadly related to the amount of vacation time that they had—in Thailand, Taiwan, and Malaysia it is typical to have fewer than ten days each year compared to over twenty in Europe—and also reflected the pace of economic change.

Call centers, where huge teams of people handle a seemingly never-ending flow of customer calls, have been described as the twenty-first-century sweatshops. With their workers referred to as battery hens and galley slaves, they are a frequent source of scare stories in the press. The call centers, which have mushroomed in the last decade with the shift toward the twenty-four-hour society demanding banking and other services at any hour of night or day, now employ over 3 million people in the U.S. Managers have threatened staff with wearing diapers if they visit the toilets too often, and one worker was disciplined for taking two six-second breaks between calls. Despite such stories, studies suggest that stress is no higher for call-center workers than others, and clearly some employers go to considerable lengths to look after their staff.[3]

But all these figures need to be taken with a large pinch of salt. Surveys can, of course, mislead, and statistics can lie, especially in an area as fluffy as stress, where the nature of the survey and the wording of the questions can heavily influence the results obtained. One particular problem is that virtually all the surveys are based on self-evaluation, making it very difficult to compare between groups of workers or between surveys. If teachers appear at the top of the stressed professions, for example, does that mean that their jobs are more stressful or just that they are more sensitive to stress? Or was the question in that survey different, prompting the higher positive response? It is also quite possible that certain personality types may be attracted to, or carefully avoid, the higher-stress/lower-stress occupations. Nonetheless, the responses to the surveys and the trends suggest that something is going on and to a sufficient extent to affect the broader economy.

One study by Sweden's Karolinska Institute, the university charged with giving out the Nobel Prize in Medicine, estimated that depression gave a reduced quality of life for 21 million people around Europe. As it costs the economy roughly 1 percent of national income, over €100 billion a year (roughly the same as $125 billion a year), it is one of the largest health-related economic problems in Europe and has implications for the redirecting of healthcare resources.[4] A study by Britain's Health and Safety Executive, the government body responsible for health and safety regulation, suggested that about half a million workers suffered from work-related stress in the latest year, the largest category after backache.[5]

Stress clearly disrupts the lives of some people who seek medical or psychological help. There is also evidence that some types of stress can be related to the development of coronary heart disease and the occurrence of heart attacks. The National Heart Foundation of Australia has, for example, concluded that there is "strong and consistent evidence of an independent causal association between depression, social isolation and lack of quality social support, and the causes of coronary heart disease," adding that the increased risk contributed by such factors is of similar order to that contributed by smoking or hypertension. It found no consistent evidence, however, for a similar association between other types of stress—acute and chronic life events, work-related stress and so-called Type A behavior or hostility.[6]

One extraordinary study showed that heart attacks rise on the days of important sporting events. A British study found that there was a 25 percent increase in heart-related hospital admissions on the day that England lost a World Cup match on penalties to Argentina in 1998. And to suggest that it is not the losing that matters, a French study revealed an increase in heart attacks in France in the days following their victory in the final of the same tournament.[7] Stress can also, it seems, lead to premature aging and shrink the areas of the brain that control memory, attention, and the ability to make decisions.

The manifestation of the stress can harm others too. According to one newspaper headline, a study showed that "A bad day at work harms your children." It advised leaving youngsters in childcare for a little longer while the parent recovers from the strains of the working day. Often, the stressed person expresses his or her frustration in some form of rage. The incidents that are picked up by the newspapers normally get generous coverage, often with good justification. The case of a Brooklyn narcotics officer who knocked a civilian unconscious during a traffic dispute and the violent Christmas-shopping stampede that killed three people in Long Island in 2008 are examples of inexplicable rage—presumably brought on by some form of stress.

People suffering from stress and anxiety while flying can display symptoms of air rage, or sky rage, which can put passengers and crew at risk. One survey suggested that 15 percent of flyers had experienced stress or air rage while in the air—this made it a more common complaint than dehydration, fatigue, or nausea. According to Syracuse University, the federal government has obtained 208 felony convictions for disruptive behavior onboard an aircraft since 2003 (though this number may be in part due to crackdowns on airline safety following 9/11).[8]

Road rage is inconsistently defined, and there is little scientific evidence available, but there is enough anecdotal information to suggest that it does occur, perhaps with increasing frequency, and does contribute to an important, if small, portion of road safety problems. One American survey of road rage (which incidentally concluded that Miami is the least courteous city for drivers, followed by Phoenix and New York) identified that stress, frustration, and bad moods, in addition to running late and being impatient, gave rise to the phenomenon.[9]

There is no simple explanation of the increase in stress levels, but many theories abound. Technology is thought to play a part. If we keep our cell phones and computers turned on, we irritate others with our devotion to the devices and feel harassed by the desire for a quick response. Yet we feel vulnerable if we switch them off! Many academics put the increased stress down to the rapid pace of change in society. The term "hurry sickness" has been coined to cover the physical and mental consequences of our increasingly pressured lives. Public-sector workers, such as teachers and nurses, are thought to suffer because they work in highly structured, hierarchical workplaces, where there is a lot of accountability and face-to-face contact with the public.

International comparisons are few and far between but can shed light on the issue. One study of English and French teachers showed that they have much in common—both cited classroom behavior, the low social status of teaching, and the lack of parental support as causes of stress. But there were also differences—22 percent of English teachers' sick leave was attributed to stress compared to just 1 percent in France, and more than half of the English teachers, as opposed to a fifth of the French sample, reported recently having considered leaving teaching.[10] In general, French teachers work shorter hours and are not required to be in school when not teaching, perhaps accounting for the differences. Surveys of teachers have also highlighted assessment inspections and the pressure to perform in exam league tables as exacerbating work-related stress.

But, even though many people feel "stress," and for some it clearly gives rise to genuine problems, others might justifiably feel that the real problem we face is a booming stress industry that needs to make us feel worse by exaggerating how bad the problem is. There is a National Stress Awareness Day held in the U.S. (it's April 16th if you're interested) and there is even an International Stress Management Association, a charity (with branches in the U.S., UK, Brazil, and elsewhere) that exists "to promote sound knowledge and best practice in the prevention and reduction of human stress."[11] A Web search will deliver a very long list of organizations and consultants that will offer individuals and businesses various solutions for their stress problems, including training in "listening skills." You can buy stress-relief sprays, aromatherapy solutions, teas, pills, and oils, which claim that "in no time, you'll feel the weight of the world leave your shoulders."[12]

Some commentators have even suggested that the word "stress" should be banned on the grounds that it is too waffly, suggestive, and unhelpful. Before using it, we should rethink and try to use a more accurate word. Do we mean *overwork, acute boredom,* or something more medical, such as *depression* or *anxiety*? Are we not in danger of making a mountain out of a molehill?

The good news is that, although the newspapers fill their pages with stories of stress and depression, they do occasionally give advice on how to cheer ourselves up. One bunch of researchers claimed a significant increase in happiness among people who followed ten simple steps. So, picking up on their recommendations, our advice is to have a laugh every day; do a good deed; treat yourself; halve your television viewing; count your blessings; say hello to a stranger; look after something you've planted; do some exercise; call or talk to a friend; and find some "talk time" to spend with your partner. If you do, you will on average be 22 percent happier in a month's time![13]

Games of Chance

"Pole 'too dangerous' for firemen"
—*Daily Telegraph*

All workplaces have their hazards, whether they be foundries, sawmills, or kitchens. Nevertheless it must have come as a shock to his fans when the *Daily Telegraph* broke the story that Noel Edmonds, a former radio disc jockey and television host tolerated by millions of Britons, was suffering from a repetitive strain injury (RSI). "Raw deal! Noel Edmonds injures his elbow lifting the telephone," the headline ran. Edmonds claimed he had contracted the injury from lifting an old-fashioned telephone used as a prop on his successful television show *Deal or No Deal*. The show draws out to inordinate length a string of blind guesses "contestants" must make to come away with a sum of money that may be one penny or $350,000 depending on the breaks. No skill is involved and the whole charade could be accomplished in moments without altering the probabilities or the outcome. But that's hardly the point. Matters are complicated by telephone interventions at various times from a "banker," who offers the contestants the chance to cut and run for some amount less than the remaining stake. This happens a dozen or so times during the course of the 45-minute show.

So Edmonds was suffering from RSI due to lifting a telephone handset weighing perhaps a pound every few minutes for half an hour or so. What are we to make of this? Should we sympathize with the $4 million-a-year presenter? Was he a genuine victim of RSI? Or was he just making a name for himself as the world's best-paid malingerer? Perhaps it was simply a bad case of inverted snobbery. For, as Edmonds confided to the newspapers, "After 40 years in entertainment, I can at last boast that I have suffered an industrial injury."

The suffering star took advice from an orthopedic consultant—a consultant with enough time between appointments, apparently, to be a "huge fan" of the daily afternoon show. He was given steroids and told to modify his hand action, which is standard treatment for the real thing.

RSI is not a condition in itself, but a catch-all term describing carpal tunnel syndrome, tendonitis, and other conditions where a repetitive action leads to cumulative physical damage to nerves, tendons, muscles or bones, usually in the upper limbs or back. Because pain is only felt gradually and there is often no visible injury, the condition can be relatively serious by the time it is recognized and require prolonged treatment. The same factors also make it easy to doubt the reality of the condition—a doubt understandably reinforced when a miraculous cure is sometimes effected by nothing more than an employer's financial settlement. Typical sufferers are not television presenters but workers at the other end of the wage scale, such as hairdressers and people processing checks. In the United States, RSI costs companies more than $20 billion in compensation claims each year, according to the Department of Labor, while in Britain the Trades Union Congress estimates that 5.4 million work days are lost.

As Edward Tenner points out in his book on technology's "revenge effect," *Why Things Bite Back*, "The human being did not evolve to perform small, rapid, repeated motions for hours on end."[1] But the contrast even between the few examples mentioned so far raises the questions: How small and how rapid are these motions? How great the strain and how oft-repeated must it be to sustain the injury? Can you get it "from opening too many bottles of wine at a Christmas party," as RSI skeptic Paul Aichroth teasingly insists?[2]

The truth is that different actions may give rise to different injuries. A single over-enthusiastic cork-pulling may cause a painful stretched muscle or a damaged tendon which will take days or weeks to heal. But this is not RSI as suffered by thousands of people in clerical jobs involving far smaller, more frequent actions. Here, the problem may lie with traditional repetitive actions such as scissoring and chopping. But with ever more widespread computer usage, it is inevitable that RSI has become identified as a technological complaint. This is made clear in Germany, where RSI is known as *Mausarm*, or mouse arm.

In the 1980s, Australia was swept by an epidemic of RSI so severe that it dented the nation's economic prosperity. Prompted by a trade union focus on occupational health in lieu of pursuing the usual wage claim, numerous measures were taken to alleviate the claimed suffering, but even now it is impossible to know whether the phenomenon had a genuine physical basis or was no more than a bad case of mass hysteria, egged on by the interest shown by the medical community. The epidemic duly tailed off, but whether due to workplace improvements or simply because there were no more susceptible people left to experience it cannot be told.

Despite spectacular outbreaks, RSI is not as prevalent as is sometimes claimed. There are several reasons for this. One is the greater recognition of this kind of workplace injury, which has led to improved design of items such as computer keyboards and the introduction of regular breaks. Another is that a proportion of those claiming to suffer from RSI are undoubtedly malingerers seeking compensation or time off work. There are, in addition, people who seize RSI as an outlet for other psychological problems such as stress. Whether they are real or imagined, RSI complaints have been almost invariably associated with employment rather than home or leisure activities.

This distinction is now becoming blurred. Many of the new technologies that are liable to give rise to RSI are personal. The new scare stories concern "BlackBerry thumb," suffered by people sending prolix e-mails on their BlackBerry or other brands of personal digital assistant (PDA), and "iPod finger," sustained from over-enthusiastic stroking of the rotary dial on Apple Computer's popular MP3 player. Cell phones may cause similar injuries in users who send too many text messages, while the compulsive sending and checking of messages is a growing addiction problem.

"Growth of PDA-related injuries a concern," *USA Today* announced, reporting warnings from the American Physical Therapy Association about improper use of the devices and opining that BlackBerry users displayed addictive behavior similar to that of smokers or alcoholics. The article noted that one hotel was offering stressed guests "a special BlackBerry Balm Hand Massage." ABC News talked of "CrackBerry addicts" who crash their cars or ruin their marriages because they can't stop using the gadgets.

Because PDAs such as BlackBerrys are often supplied for work, some employees are beginning to sue their employers when these complaints arise. But with iPods and mobile phones, users generally have nobody but themselves to blame. "iPod finger" made its debut as a medical condition following a warning by the British Chiropractic Association in late 2005. Closer inspection of the story, however, revealed an admission by the association's Dr. Carl Irwin that few actual injuries had yet been reported—but it was "only a matter of time."[3]

"Millions more" are at risk from texting, according to a *Daily Express* article addressed to the nation's "text-maniacs." The British Chiropractic Association had already warned in 2004 of the dangers of excessive text messaging. Apparently without irony, it suggested that you might avoid problems if you were to "shrug your shoulders" before and after texting.

Ever since Pac-Man and Space Invaders, game consoles have been another focus of concern because of both the repetitive actions they demand and the sometimes addictive nature of the games. As the leading manufacturer of such games, Nintendo became associated with an RSI condition inevitably known as "nintendonitis." Nintendo may have been unfairly singled out, as a 1998 Cornell University study suggested that increased use of computers in classrooms was a similar liability. Nonetheless, Nintendo's games now carry copious instructions to rest the hands, take breaks and so on, but, it seems, sometimes to little avail. Apparently, extravagant actions by players of Nintendo's new Wii game console combined with a loss of grip at the crucial moment have led to the units flying across rooms and striking priceless vases and grandparents in their path. Enthusiasts have posted images of their "wiinjuries" on the Internet, and news sources ranging from the *Boston Herald* to NPR have found cause for half-joking, half-concerned reportage on the issue. Nintendo responded by offering users more robust versions of the wrist straps supplied with the game.

The injuries occasioned by these personal hi-tech gadgets may be just as real as those sustained from poorly designed equipment in the workplace, but they are often borne with more fortitude because of the cachet the user gains from possessing the object. Thus, while the devices proliferate, the government is not too quick to intervene.

But you don't even need to use these devices to be at risk, according to some sources. "Cell phones increase risk of death by lightning, doctors claim" read a *LiveScience* headline picking up on a report from three doctors in the *British Medical Journal* of a girl hit by lightning while making a call in a city park. According to London's *Metro*, the lightning provoked a heart attack, and the girl "was found lying on her back. One arm was stuck rigidly upward and her burnt-out phone was clutched in her hand." Fortunately, she escaped complete technological martyrization, surviving with injuries including the loss of hearing "in the ear she used to listen to her [cell phone]."

The *Metro* article went on to mention three phone users in China, Korea, and Malaysia described in the *BMJ* piece who had been struck dead by lightning. The good doctors wrote that "education is necessary to highlight the risk of using mobile phones outdoors during stormy weather," but they said nothing about the risk of wielding non-headline-grabbing metallic objects such as umbrellas. Nor was any attempt made to establish whether the risk of injury or death was materially increased by carrying a phone. Deaths due to lightning strike are surprisingly high. The National Weather Service puts the average U.S. death toll due to lightning at seventy-three people a year; the global figure must be over a thousand. Given the high ownership of personal technology, it is remarkable that the *BMJ* authors could find only three cases world-wide where the victims were using cell phones.

How confusing all this is. Is it a disgrace that people are allowed out in storms without realizing the danger? Or ought we to defend our right to roam in all weathers from the do-gooders? Newspapers have recently carried reports that schools have banned children from playing schoolyard games, such as soccer and touch football, that drivers have been prevented from parking their cars under fruit trees, that firemen may no longer slide down poles, that cats may no longer be rescued from trees, and that hanging flower baskets must no longer hang if they do so over people's heads, all these things supposedly decreed by zealous health and safety officials.

If a single person were seriously hurt or killed by any one of these curious hazards, the newspapers would naturally be eager to report the story. But in the meantime the media stance is one of pious disparagement of the people whom the *Daily Mail* terms the "health and safety killjoys." Perhaps things *have* gone too far in the name of risk reduction. A new chairman of Britain's Health and Safety Commission began his term with an unexpected attack on the pedants in his industry. His advice, passed on in a *Daily Telegraph* headline, was for people to "Get a life and take sensible risks." Okay. Now what, exactly, is a *sensible* risk?

6. Law and Order

Since 9/11, terror has become a big concern for ordinary people in ordinary places and everyone has a view on the merits of the "war on terror." But just how scared should we be of a terrorist attack? And can a government ever pitch its advice at the right level without scaremongering or appearing to ignore the risks? What are the consequences of real wars? Traditional crime, however, has more day-to-day impact on people and stories about murderers roaming our streets seem designed to scare. The early release of criminals from prisons adds to the worry.

Terror Alert

"U.S. raises terror alert level"
—*The New York Times*

Planning to see the Pyramids? If so, the travel advice on the Web pages of the U.S. Department of State will probably put you off: "In September 2008, 11 foreign tourists and 8 Egyptians were kidnapped for ransom in the remote south-western desert region, close to the Sudanese border. Tourists should avoid travel to the border region." It doesn't sound very promising, does it? But then again, there is similarly gloomy advice for other places, such as Morocco or Turkey, where people seek some winter sun in their thousands. The situation is worse in other potential holiday destinations such as Thailand and Indonesia (including Bali), and in both cases the Department of State advice omits the reassuring phrase "Most visits are trouble-free."

Is the situation that bad? Is the Department of State erring on the side of caution? It does not warn against travel to every country where there is a risk of terrorists operating—for the obvious reason that it would cover virtually the whole world, and also, only somewhat reassuringly, because many terrorist groups "may not be looking for American targets." And there are countries to which it advises against all travel, like Somalia. It also advises against all travel to parts of another 25 or so countries—including India, Indonesia, Israel, Pakistan, Russia, and Sri Lanka. That said, although detailed figures are not given on the Web site, the reality is that in countries like Egypt and Turkey far more visitors die from natural causes (like heart attacks among the elderly), drowning, road and rail accidents, and natural disasters, such as earthquakes, than from terrorism.

The Department of State advice reflects—or perhaps along with other government agencies operating in the domestic environment stirs up—the sense of fear in which a large proportion of the developed world lives. During the time of writing, the Department of Homeland Security assessed the security threat to the U.S. from terrorism as being "elevated" and the risk on domestic and international flights as being "high." The latter is the fourth point on a five-point scale, going from low, guarded, elevated, high, and finally severe. Vaguely enough, "elevated" meants "significant risk of terrorist attacks," while "high" means "high risk of terrorist attacks." You wouldn't guess from watching the people wander the streets that we were on as high a level of alert as possible for that length of time without the authorities having knowledge of an imminent attack. Apart from asking all Americans to be "vigilant," it scarily said that "everybody should establish an emergency preparedness kit as well as a communications plan for themselves and their family."

Such a level of concern is reflected in the many alternative surveys that have been conducted assessing perceptions of the terrorist threat. One study showed terrorism to be ahead of all other worries in both the U.S. and Europe, with 97 and 94 percent respectively of those polled believing it to be an important threat.[1] Another survey showed that 90 percent of Britons and 84 percent of Americans felt that their country is likely to be the target of a terrorist attack in the near future.[2] A third put the likelihood of an attack a little lower—about two-thirds of respondents in the 2 countries thinking that an attack was somewhat or very likely in the next 12 months.[3] Of course, the response depends on the exact wording of the question that is asked, but it is clear that concerns are there and they seem to be affecting our daily behavior—roughly 6 out of 10 Americans and Britons now look twice at other passengers on public transport.

But the figures are not high in every country. While the levels of fear in France (84 percent) and India (82 percent) are close to those seen in the U.S. and the UK, other countries are lower—for example, Germany (47 percent)—and some are much lower—for example, Hong Kong (12 percent) and Hungary (17 percent). Worries about terrorism are highest in countries that have already suffered an attack. It is only those countries that have to date been spared attacks that have citizens with greater worries than terrorism—for example, the dominant concerns in Hong Kong are SARS and bird flu.

Furthermore, concern about terrorism is not evenly spread within countries. One American survey showed that nearly a third of those who lived in big cities said they were "personally very concerned" about an attack, compared to only one in eight of those living in small towns or rural areas.[4] In New York, 69 percent of people are "very concerned" about another terrorist event, only fractionally down on the percentage recorded in October 2001 in the wake of the twin towers attack. The concerns of New Yorkers about terrorism remain high, but across much of the U.S. concerns diminish as the vividness of the memory of the events of 9/11 fades. As if the fear of the terror act itself was not enough, surveys tend to find that a majority of residents in developed western countries feel that their country is not well prepared to respond to a terror attack.

Even economists are getting in on the act of worrying about terrorism: A prestigious survey of America's dismal scientists showed that they believed the threat of a terror attack was the biggest short-term problem facing the U.S. economy.[5] Most people think that the failure to find a successful conclusion to the Iraq war has increased rather than decreased the perceived chances of terrorist attacks. One large survey spanning 35 countries found that 60 percent of those polled shared this view, with only about one-quarter believing that the war had reduced the chances of an attack or had had no effect. In Britain, in contrast to the government view, 77 percent of those questioned thought that the terrorist threat had risen since the 2003 invasion.[6]

The fear has had some odd consequences. One of Berlin's top opera houses came under fire for canceling a controversial production of Mozart's *Idomeneo*, which was to show the severed heads of the Prophet Muhammad, Jesus, and Buddha, due to concerns about Islamist attacks. German politicians condemned the cancelation as self-censorship, cowardice, and damaging to the principle of free speech. The decision came in the wake of the riots following the publication of cartoons of Muhammad in a Danish newspaper. There have been reports of other, seemingly absurd, reactions in Germany designed to avoid offending the 3.5 million Muslim residents. These allegedly include a request to darken a school gymnasium when Muslim girls are working out and a woman being told to change the name of her horse from "Muhammad" to something less troubling, such as "Momi."

Much of the concern of westerners is focused on the concept of "a Muslim problem." One poll in Britain suggested that 53 percent of people felt threatened by Islam, a rise from 32 percent in the wake of the September 11 terrorist attacks on America. The appearance of a so-called clash of civilizations is supported by the fact that there has been a near doubling, to one in five, in the proportion of Britons who believe that "a large proportion of British Muslims feel no sense of loyalty to this country and are prepared to condone or even carry out acts of terrorism."[7]

But just how real is the threat? Some of the headline figures are, indeed, pretty scary. The roll call for global terrorism in 2007, compiled by the National Counterterrorism Center in the U.S., included over 14,000 terrorist incidents and 22,000 deaths of noncombatants, with many more thousands wounded and kidnapped.[8] The figures are high enough, but the trend is even worse: The figures for 2006 were well up on those in 2005, which in turn were a hefty increase on the previous year. Part of the increase in 2005 is explained by definitional changes to the figures. In the light of the American government's increased focus on the threat from terrorism, the authorities decided that the State Department's annual publication needed to be changed, effectively ending a twenty-year run of consistent data.

The definition of "international terrorism" was shifted from one of "involving citizens or territory of more than one country" to a new definition of "terrorism" meaning "premeditated, politically motivated violence perpetrated against non-combatant targets by sub-national groups or clandestine agents, usually intended to influence an audience." This might sound trivial but it has made a large change to the figures. As the new definition has been brought in, the figures have increased. Some government critics cried foul, suggesting that the changes were only made to make the numbers larger, making it easier for the government to justify its policies. The old figures were criticized, however, as they failed to count some clear cases of terrorism, including the Van Gogh assassination, the Philippine super ferry and the blowing-up of a Russian aircraft. While a simple aggregation of terrorism incidents is hardly a perfect metric for measuring the scale of the problem or the success in tackling it, such a list does provide a useful basis for analysis.

The figures show that the Near East and South Asia are particularly hard hit by terrorism, accounting for 75 percent of the attacks and 85 percent of the fatalities in 2007. Nearly two-thirds of the non-combatant fatalities worldwide in 2007 occurred in Iraq, with India, Afghanistan, Sudan, and Sri Lanka collectively accounting for another large proportion. Several categories of civilians, including police, religious figures, and journalists, bore a disproportionate brunt of terrorism. Over half the victims killed or seriously injured in 2006 were Muslims. Away from Iraq, Afghanistan, and a handful of other mainly African terror hotspots, the number of reported terror incidents fell in 2007, continuing the recent trend.

The ex-Iraq numbers are not that large. This is reflected in the detailed information of attacks on the Web site of the UK's Foreign Office—it lists only a couple of dozen terrorist incidents from the previous three years, focusing mainly on incidents in generally terrorist-free areas. A little over 500,000 people die each year in England and Wales yet typically only a dozen or two Britons have died as a result of terrorism in recent years. Annual deaths from terrorism have been much lower than deaths from transport accidents (3,000), falls (3,000), drowning (200), poisoning (900), and suicide (over 3,000).[9] Twenty-eight non-combatant U.S. citizens were killed by terrorists in 2006, of which 22 were in Iraq and 3 in Afghanistan. It is pretty clear that, as long as you stay away from the world's insurgent hotspots, the chances of being caught up in a terrorist event are minuscule.

But to what extent can we ignore the threat? One view is that the reaction of western governments to the terrorist threat is exaggerated. Terrorist attacks can never amount to more than a "big, bloody nuisance," as one columnist put it—and we should carry on with our lives as normal. An alternative view is that the government should do more, introduce new security measures and make more arrests, in order to minimize the possibility of further attacks.

The low fatality figures quoted above would appear to support the former argument. It has been pointed out that in most years, more people die in their baths than at the hands of international terrorists, and that more Americans have been killed by lightning or an allergic reaction to peanuts than by terrorists. The argument continues that the cost of hasty and ill-considered anti-terror measures and the impact of panic on liberal institutions are far greater than the cost of the terrorism itself. But even a convincing argument against panic does not mean that there is no need for a vigorous policy to combat terrorism. In terrorism, as in crime more generally, the number of incidents is contained by the policies in place to combat it. Terrorism might after all be like a disease that starts with only a few people and becomes an epidemic, at which point it gets out of hand. Perhaps we should consider ourselves fortunate that terrorists are, with some notable exceptions, generally so inept.

The statisticians are doing more than publishing figures; they are playing a role in maintaining safety levels in society by using their skills in risk analysis, profiling, and screening, coupled with the rapid identification of disease clusters in the event of biological or chemical attack. Meanwhile we are left with the near-impossible task of judging the tiny risk of an apocalyptic event. With the British newspapers saying in the summer of 2006 that there are at least 1,200 home-grown Islamic fanatics under surveillance by the security services in Britain, it is no wonder that many people remain nervous. If that weren't enough to worry about, by the end of 2006 the British counter-intelligence and security agency was saying that there were 1,600 "terror plotters" up to no good in the country[10] and by February 2007 the number had risen to "more than 2,000."[11] That "inflation rate" is certainly enough to get the pulse racing.

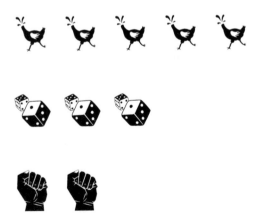

Bang Bang

"Disputed study: Over 650,000 civilians killed in Iraq war"
—*Fox News*

Television images and newspaper reports give the impression that violent conflicts and uprisings of one sort or another are sprouting up across the world. Yet the people who keep tally of such things reckon that the last two or three years have seen the number of wars drop to the lowest level since the mid-1960s. It seems that only about half the number is taking place now compared to the peak in the early 1990s—some twenty-eight wars were counted in 2006 compared to fifty-five in 1992.[1]

Notwithstanding all the methodological difficulties in counting and classifying such activities, the downward trend in the last decade or so represents a stunning turnaround from the seemingly inexorable rise over the preceding twenty years. Data are also collected for "armed conflicts" which are of lesser severity or which for other reasons do not qualify as full-scale wars. These have also been on a downward trend in recent years although the data have only been collected for a decade or so. A different source collects figures on the number of "political conflicts" and assessed that there are around 270 live, of which 10 percent involved a high level of armed violence and 30 percent involved occasional violence.[2]

Although the number of conflicts might be declining, very little is known about how many people fall victim to this organized violence. Most data collection efforts are focused on battle deaths, ignoring civilian deaths, especially those which may result from the indirect consequences, such as failing economies and collapsing health systems. The second-round effects of conflicts are laid bare in the example of the Democratic Republic of the Congo. A study in Britain's *Lancet* estimated that nearly 4 million people have died there since 1998, and that less than 2 percent of the deaths were the direct result of violence, with most people succumbing to disease and hunger.

The difficulties of counting war dead came to a head in October 2006 when the *Lancet* (again) published a controversial study regarding the number of dead in Iraq since the U.S.-led invasion in March 2003.[3] It suggested that more than 600,000 people had been killed since 2003, equivalent to 2½ percent of the population, and much higher than other estimates derived from counting bodies in mortuaries or tabulating media reports of deaths. The widely quoted Iraq Body Count, one such compilation of media reports, gave a death toll of about 50,000 over the same period.[4] Experts in the field typically say that body-count methods underestimate the true count by a factor of at least five, but there has been no "scientific" corroboration of this.

Although President Bush, who described the estimates as "not credible," along with the U.S. commander in Iraq and Iraqi officials questioned the methodology, the figures seemed to confirm the widely held impression that Iraq was descending into what the researchers called "blood-thirsty chaos." They surveyed nearly 2,000 households at 47 sites across Iraq, asking about births, deaths, and migration in and out of the areas. The mortality rate had more than doubled since the invasion, and more deaths, which were mostly confirmed by death certificates, had been caused by gunshot wounds than anything else.

Arguing about the exact number, which will never be known, is to miss the point. In any case, the researchers did give an admittedly wide range of the estimate of the number of deaths from 390,000 to 940,000. Rather, it is the scale of the carnage hinted at by the survey that is important. Even a figure close to the bottom of the *Lancet* range would be roughly ten times as great as the "official" estimate of civilian deaths—30,000 to 50,000 was typical of the figure coming at the time from the administrations involved in the war. Yet even the low figures have the power to shock: One estimate, 50,100, of the number of Iraqi dead made in the summer of 2006 prior to the *Lancet* study was described in one newspaper as "staggering."

Needless to say, such figures of Iraqi dead are very politically sensitive. The U.S. general Tommy Franks was famously quoted as saying, "We don't do body counts," and, at least until late 2005, the Pentagon claimed that it did not keep a running total of Iraqi deaths. It refuses to make public the figures that are collected, presumably for fear of inflaming concerns about the war. The sensitivities were emphasized by the news in October 2006 that a leaked memo from the office of the Iraqi prime minister to the country's Health Ministry had asked that it stop providing mortality figures to the United Nations.

Deaths among the Allied troops are, in contrast, logged in meticulous detail. By mid-2008, nearly 4,500 coalition troops had been killed in Iraq: over 90 percent were Americans and the majority of the remainder British. The lists of servicemen deaths on the U.S. Department of Defense and British Ministry of Defence Web sites are collated on other sites run by interested parties. The coalition has so far been far less enthusiastic when it comes to publishing either death rates (the number of dead relative to the number serving) for its military personnel or the number wounded. While any individual death is regrettable for those immediately involved, from the point of view of the media, public opinion, government policy, and military planning, it is arguably more important to have the death rates that measure the risk of death for personnel located in any particular area.

Some independent analysts have tried to compile such figures from official sources, but this can be difficult without being sure how many forces of various types are stationed in each country at any one time. One such study suggested that the death rate (the number of deaths per thousand years of serving personnel) of British soldiers based in Afghanistan was six times higher than for their peers in Iraq in 2006, double that for those who were involved in the initial invasion of Iraq in 2003.[5] The author of the report said that "The commentary we are getting from politicians about the conflicts does not do justice to the threat our forces face." The authorities resisted demands for the publication of figures showing the extent of injuries to the forces on active duty.

The pattern of declared military expenditure by the world's governments has followed a slightly different path from that of conflicts. Expenditures (in inflation-adjusted terms) peaked in the early 1980s and, after a period of stability, fell in the early 1990s at the end of the Cold War. By the mid-1990s, expenditure was only two-thirds of the level in the late 1980s. More recently, however, and especially since the September 2001 attacks in the U.S., military spending has been rising more strongly. Countries in many regions of the world have been increasing their expenditure, but the rises have been more modest in Central and South America and in western Europe, notably in Brazil, Mexico, and Germany, where expenditure has been cut back.

The U.S. spends almost as much on military activities as the rest of the world combined. The country's spending is at its highest level since 1969 even though the number of armed forces (soldiers and civilians) stands at around 2 million, less than half the number forty years ago. Employment in the arms industry is close to 4 million, a million more than it was at the height of the Vietnam conflict. After the U.S., the big spenders are the UK, France, Japan, and China—together they account for a further one-fifth of global expenditures. The top 15 budgets together account for over 80 percent of the global total, with NATO countries being the dominant spenders. The large budgets of Russia, China, and India are difficult to assess accurately due to both a lack of transparency in public and corporate spending and the difficulty of choosing an appropriate exchange rate for the purposes of comparison.

Per capita annual expenditures vary enormously between countries, ranging from nearly $1,500 in North America, and $540 in western Europe, down to just $18 in Africa. It has been estimated that developed countries spend roughly ten times as much on the military as they do on development aid for the poorer countries[6]—a ratio that many people find hard to understand when large chunks of the world are suffering from poverty and ill-health.

More positively, the spending by the United Nations on peace-keeping operations is now at a record high, reaching just over

$5 billion in the year to mid-2006, roughly double the amount at the turn of the millennium and surpassing the previous peak of 1994. Roughly 70,000 soldiers, police, and military observers were serving in 16 peacekeeping operations at the end of 2005. Over 40 percent of the peacekeepers are located in Pakistan, Bangladesh, India, Nepal, and Sri Lanka, with 6 sub-Saharan African countries accounting for another 20 percent. A good part of the increase in budgets in the last year is accounted for by the missions to Congo and Sudan.

When servicemen return from active duty you would think that their problems were behind them, but many media reports suggest that is not the case. Some stories suggest that soldiers have difficulty getting appropriate medical treatment, and that the failure to give them priority means that thousands are unable to return to the front line. It is quite possible, however, that the psychological consequences of fighting in these wars will be greater than the physical consequences—between 15 and 20 percent of American servicemen in Iraq show signs of depression or post-traumatic stress disorder.[7]

One particular problem facing ex-servicemen is the so-called Gulf War syndrome, referred to by one newspaper as the "mystery plague that wrecks 8,000 lives." The existence and nature of the syndrome is a subject of considerable debate as, while there is incontrovertible evidence that rates of ill-health are greater for Gulf War veterans than across the armed forces in general, there is no evidence of an increase in mortality or the occurrence of any well-defined medical condition.[8] Symptoms attributed to the syndrome have been wide-ranging, including chronic fatigue, loss of muscle control, migraines, dizziness and the loss of balance, memory problems, muscle and joint pain, sexual problems, indigestion, skin problems, mood swings, and shortness of breath. The British and American ex-troops have the highest rate of excess illness and are distinguished from the other nations by higher rates of pesticide use, use of anthrax vaccine and somewhat higher rates of exposure to oil-fire smoke and chemical alerts. Among the most likely potential causes not yet excluded are combustion products from depleted uranium munitions, side effects from anthrax and other vaccines, parasitic infectious diseases, the spraying of insecticides over tents, and exposure to chemical weapons such as nerve gas or mustard gas.

The combination of multiple possible causes, multiple symptoms and the background of higher reported cases of stress disorder means that we are unlikely ever to have clarity on the story. That was certainly the conclusion of fifteen years' research by scientists from the Royal Society, the British science academy, who concluded that there was "little value in conducting further research."[9] This will be of little reassurance to those veterans of the current Iraq war who are already revealing a range of serious health issues.

Crime and Punishment

"Sex offender worries neighbor"
—*Seattle Times*

The safety of you and your family is paramount, and there are few things that are more scary than thinking you are under threat from dangerous souls wandering the streets. Every day, it seems, the media will have another tragic story about an innocent person who was in the wrong place at the wrong time and met an untimely end. The sight of bunches of flowers being tied to railings at the scene of a murder is a moving and seemingly frequent occurrence on our news broadcasts. Crime stories are also strongly political. The hyping-up of crime figures, selecting those that make the situation look worse, is a tried and trusted way of making any government look bad.

The crime rate in the U.S. is not, as some people believe, static. New York recently saw the murder rate drop to its lowest since 1963 on the back of the "zero tolerance" policies pursued from the mid-1990s. New York is now the safest big city in America, while others, such as Boston, Houston, and Philadelphia, see rising crime rates. The Houston murder rate rose by one-fifth in the year when 150,000 evacuees arrived following Hurricane Katrina.

Still, the chance of someone being involved in gun crime is very small as it is largely confined to small pockets of inner cities—most of the shootings happen in a handful of the nation's poorest cities. Mostly the disputes are about drugs, and three-quarters of shooting victims are young and black. Both the homicide victimization rates and the offending rates for blacks are six to seven times higher than for whites. Statistics setting out the likelihood of different types of people dying as a result of homicide are readily available in America. One Web site coldly sets out the factoids: the 21.2 per 100,000 black people who died from homicide in the stated year compared with a rate of 8.3 for Hispanics, 4.9 for whites, and 3.3 for all women.[1] As 85 percent of murder victims know their killer, it is pretty clear that, unless you are keeping some bad company, you're pretty safe.[2]

But what becomes of the "bad company"? In the U.S., the rate of incarceration has more than trebled in the last 25 years. It is now just short of 500 prisoners per 100,000 population, meaning that one in every 200 people is in prison. Over 2.1 million Americans are in jail (pending trial or serving short sentences) or prison compared to under 1.5 million a decade before.

Whatever the explanation for the trends, the building of prison spaces has failed to keep up with demand. This has led to overcrowding, with occupancy levels, based on official capacity, at over 100 percent. The overcrowding puts pressure on prison management and disrupts the smooth running of prison regimes, raising the threat of rioting and higher rates of prisoner suicide.

The increased churn of prisoners interrupts training programs and undermines education, training, and rehabilitation. But such help is vital due to the profile of the prison population—prisoners are much more likely than the general population to have run away from home as a child, have no qualifications, suffer from two or more mental disorders, and be drug users.

Prison overcrowding panics people because it means that an increasing number of offenders are released back into the community before the end of their term, possibly to reoffend. In response to a similar problem in the UK, an article entitled "Number of freed lifers trebles in five years" quoted a member of parliament saying, "The public are being failed by a system which allows murderers and rapists back onto the streets to commit more offences. A life sentence should mean what it says, but at the moment it just means a few years watching television in a comfortable cell."[3]

In recent years, several high-profile murders have indeed been carried out by criminals on probation, showing that it is difficult for the stretched professionals to assess accurately the danger posed by offenders. In the U.S., three-fifths of convicts from state prisons fail successfully to complete their parole. Governments have never found it easy to get the appropriate level of policy coordination among the various groups involved—police, courts, prisons, and probation services.

In response, a large number of Americans—especially parents of young children—have demanded safeguards against becoming an ex-con's unwitting next victim. Search engines like Family Watchdog and Map Sex Offenders allow visitors to find sex offenders who live in their area—in case one is in that visitor's very own neighborhood. However, such alert systems may in themselves be contributing to the crime problem, as cases like one in Los Angeles in 2005—in which two known sex offenders were shot and killed in their home—might suggest.

If anything, we might expect crime to play an even greater role in the future in media-induced panic. More and more countries are coming under pressure to introduce national crime figures broken down by town or administrative area. Only then will residents have

Crimes reported by the police, 2001

	Homicides	Homicides per 100,000 population
Australia	340	1.9
Canada	554	1.8
England and Wales	891	1.6
France	1047	1.7
Germany	868	1.1
Ireland	59	1.4
Italy	818	1.5
Japan	1340	1.1
Russia	33583	22.1
Scotland	107	2.2
South Africa	21683	55.9
Spain	494	1.1
U.S.	15980	5.6

Source: International Comparisons of Criminal Justice Statistics 2001, Issue 12/03, October 24, 2003, www.homeoffice.gov.uk. Some data relate to 2000.

a better idea of the risks that they face and will police and others charged with keeping law and order be able to best direct their scarce resources. As more data become available, more stories will be written. In America, many of the newspapers keep a close eye on how the crime figures are moving. "Keeping monthly homicide statistics can be like watching a horse race. They're ahead. They're behind. They're neck and neck. But there are no winners. Only losers."[4]

7. The Natural World

Early civilizations believed in a world ruled by angry gods with the power to unleash forces of destruction beyond our control. Today, we understand that these forces are natural and not divine. But for all our science, we still have a hard time predicting or controlling them. To make matters worse, the scale of human activity on the planet is now such that we are able to whip up global disasters all by ourselves—though perhaps also to prevent them.

Looking Up

"Ozone layer healing, but more slowly than hoped"
—*Washington Post*

We have grown up worrying about the hole in the ozone layer. From the 1970s, depletion of this fragile layer of the Earth's atmosphere by man-made chemicals called *chlorofluorocarbons* (CFCs) had been forecast to expose humanity to unprecedented levels of dangerous solar radiation. Strong ultraviolet sunlight would penetrate the atmosphere, reaching the ground, where it would cause skin cancers and eye damage in humans, livestock and wildlife and impair the photosynthesis of plants. "What will it do to our children's outlook on life if we have to teach them to be afraid to look up?" Al Gore wondered in his first essay on environmental apocalypse, "Earth in the Balance."[1] Greenpeace ran an advertisement headlined "GOODBYE SUNSHINE," showing a photograph of a baby shielding its eyes against the "harmful radiation reaching the Earth's surface."

Children are still able to look up. So what went wrong, or rather, right?

Ozone was once regarded as a health-giving substance. In Victorian times, it was used in the way that chlorine is now for cleaning, bleaching, purifying water, and preserving foods. People were sent to the seaside for their health, where the air was said to be rich in ozone, although you are more likely to catch a whiff of it near high-voltage electrical equipment, which can convert small amounts of the diatomic oxygen that we breathe into triatomic molecules of ozone.

The ozone layer had been known for more than fifty years when it first became a matter of environmental concern at the beginning of the 1970s. It lies in the stratosphere, the dry layer of the atmosphere between 6 and 30 miles altitude immediately above the troposphere. Whereas the troposphere is dominated by physical processes we experience as weather, the action in the stratosphere is chemical. The fear at this time was that exhaust chemicals from supersonic aircraft flying at stratospheric altitudes would destroy the ozone, which blocks strong ultraviolet light from reaching the Earth.

Simultaneously, scientists began to realize that ultraviolet light might be a factor in causing cancer. The coincidence was too great to ignore, and both the danger of ultraviolet exposure and the potential dangers of fleets of supersonic aircraft crisscrossing the skies made the headlines. Any substantial threat was quickly nullified, however, when the United States Congress halted Boeing's supersonic passenger plane development project, although the reasons for this were ultimately economic.

Meanwhile, the environmental scientist James Lovelock, later to become well known for the Gaia theory, had used a sensitive device of his own invention to detect minute concentrations of CFCs in the atmosphere. There was no other source for these chemicals other than the refrigerants, aerosol gases, and foaming agents that had begun to be manufactured during the Second World War. A little later, scientists at the U.S. National Center for Atmospheric Research discovered that chlorine atoms catalyze the breakdown of ozone in the atmosphere. Lovelock's gadget had been built to monitor pesticide levels, while the NCAR team was concerned about chlorine spewed into the atmosphere by erupting volcanoes. Nobody suspected a link between ozone depletion and CFCs.

But in 1974, two chemists at the University of California, Sherwood Rowland and Mario Molina, established that CFCs were the dominant source of chlorine that was indeed breaking down ozone in the stratosphere. The discovery was startling. It showed that the chemicals we had started putting into refrigerators, spray cans, and air conditioners 20 years before could escape (due to inefficient factory processes and poor disposal practices) and drift lazily up through the atmosphere, taking years or even decades to reach the ozone layer. Even at concentrations of less than one part per trillion, these chemicals are there broken down into chlorine atoms, each one of which can take out tens of thousands of ozone molecules in a vicious cycle of destruction. When it appeared in *Nature*, Rowland and Molina's paper, however, was not one of those highlighted by the magazine's press office, and it was ignored by the world's journalists.

The attitude of the media changed abruptly in 1985, when Joe Farman of the British Antarctic Survey reported a significant loss of ozone above the South Pole. Both the extent of the depletion and its latitude came as a surprise to scientists, whose models had led them to expect to see the first evidence of ozone depletion in the upper stratosphere above the tropics. When Farman's findings were confirmed by NASA, the *Washington Post* reported the discovery and coined the phrase "ozone hole." No matter that it described the seasonal, partial depletion of a very minor atmospheric constituent and the consequently increased permeability of the atmosphere to merely part of the sunlight spectrum, the "hole" had an undeniable graphic appeal. It was a rent in the firmament that would allow evil ultraviolet light to pour in, withering our crops and mutilating our bodies.

The advent of the ozone hole was an undoubted spur to those negotiating the Montreal Protocol to limit CFC production, although discussions had begun as early as 1981. When the protocol was signed on September 16, 1987, 24 countries agreed merely to level off and then gently reduce their output, not to outright ban them. Yet Americans had legislated a decade before to ban the use of CFCs as aerosol propellants in response to the research into the likely effects of supersonic aircraft—and American consumers had actually done much of the work first by refusing to buy them in any case. This, and not Montreal, was "the first, and last, unequivocal application of the precautionary principle in the ozone story" according to Joe Farman.[2]

The idea that Montreal would fix anything was slow to take off. In 1988, after the protocol was signed, the science writer John Gribbin published *The Hole in the Sky*, a popular account of "man's threat to the ozone layer." "This book is required reading for everyone who is concerned about the gigantic threat poised over us all," trumpeted the cover blurb. Gribbin was convinced that Montreal's half-hearted measures wouldn't be enough. "Tomorrow is too late," he wailed.[3] In fact, as Gribbin of course knew better than most, tomorrow would do pretty well because of the 20-year delay built into the migration of CFCs up into the stratosphere.

The ineffectual original agreement was gradually strengthened at a sequence of later international gatherings, as more countries signed up, including the major developing countries, and reduction targets were made more onerous. Developed countries finally stopped making CFCs in 1993, several years earlier than originally planned. China is due to phase out production by 2010. In 1995, Rowland and Molina shared the Nobel Prize. The Swedish Academy's citation gushed that the researchers had "contributed to our salvation from a global environmental problem that could have [had] catastrophic consequences."

Not so fast. Even in the 1990s, it was not yet clear that the CFC deals would produce the desired effect. In 2002, NASA predicted that repair of the Antarctic hole would only be *detectable* by 2020. Annual monitoring since then has revealed no clear trend and suggests that atmospheric systems are more complex than scientists had hoped. The formation of the hole after each Antarctic winter is subject to disruption. The hole shrank significantly from 2003 to 2004 but grew in 2005 and then again in 2006. Temperatures in September 2005 were the lowest since 1979, accelerating the seasonal formation of a hole which in that year peaked at 10 million square miles. But the hole also healed again during October, earlier than in previous years, which scientists attribute to changes in local meteorological conditions, not to changes in levels of ozone-depleting chemicals. Such blips, according to the World Meteorological Office,

> cannot be explained by changes in the stratospheric halogen loading, but are due to interannual dynamical variability. This variability will make it difficult to detect the onset of ozone recovery in Antarctica, and in particular it will be difficult to attribute any positive change in ozone to declining amounts of ozone depleting substances.[4]

The latest measurements offer a little more hope. The U.S. National Oceanic and Atmospheric Administration announced on August 23, 2006 that the Antarctic hole had stopped expanding and is expected to have healed by around 2060. "Healing" is a relative term, of course, just as the word "hole" marked an arbitrary threshold of ozone loss. Joe Farman is less optimistic. Although very little chlorine-containing gas is now being released, what's there already can only decay very slowly. "One third of it will still be there in 2100: that's an awful lot of chlorine."[5]

The ozone hole clearly retains considerable scare value. Even the generation that has grown to adulthood since the Montreal Protocol was signed frets about it. Their fears may be seriously off target—two-thirds of a sample of people in one survey apparently believed that the ozone hole is causing climate change.[6] But the fact that they persist indicates that the ozone hole has become an icon. It offers a ready image of man-made environmental degradation that increased emission of carbon dioxide does not.

So it would be a shame to let it go to waste. Fortunately for the environmental Cassandras, there is still the Arctic to worry about. Although NOAA's latest estimate is that the smaller and less frequently appearing hole here will be repaired as soon as 2030, some scientists fear that it will continue to grow. A large Arctic ozone hole would lie far closer to major centers of population than the Antarctic hole, and pose a correspondingly greater danger to human health, exposing more than 700 million people to dangerous levels of ultraviolet radiation.

How can this threat be real if we have largely halted the release of CFCs? The potential difficulty this time is with natural chemical processes triggered by abnormally low temperatures in the stratosphere—due, perversely, to global warming affecting the atmosphere lower down. According to the environment writer Fred Pearce, the Arctic ozone layer is now on "a hair trigger," with "many researchers" expecting a "giant ozone hole to form" as it did a generation ago over the Antarctic.[7] But the hole is not there yet. Predictions of a large hole for 2005 were later said to have been "overstated." The next year did see some worsening in the situation. But then in 2007, the hole shrank by 30 percent. The folly here is to expect to see in each year's tidings a clear sign of what will happen in the future, when the overall picture is one where any long-term trend must be picked out from amid a chaos of short-term variations. Succumbing to this temptation has led to some silly analysis not only by the media but also by partisan scientists and sociologists of science overreacting to a given year's new data.

In case a new hole doesn't appear—and the mechanisms of these processes are poorly understood at present—Pearce has another message of doom. Suppose we do successfully avoid or repair both ozone holes. This would once more ensure that less ultraviolet light reaches the lower atmosphere. This in turn would mean reduced production of another atmospheric constituent, hydroxyl, which mops up carbon monoxide and other pollutants. All of a sudden, it's a problem to fix the problem!

The ozone episode may or may not be over. So what are, or were, the costs? At the time of Montreal, the U.S. Environmental Protection Agency predicted a 5 percent rise in non-malignant melanomas for each 1 percent loss of ozone, amounting to perhaps 20,000 extra American victims a year. The prediction for the life-threatening malignant melanomas was not so clear. Al Gore passed on reports that Patagonian fishermen were catching blind salmon. Yet others found there had been no "demonstrated harm to people or other living things" by 1990.[8]

Ozone depletion has continued since then until possibly stabilizing today, and melanoma rates have indeed risen widely. The effects of ultraviolet radiation on the body are better understood than they were, but the delayed emergence of melanomas means that it is hard to establish their cause. Australia has the highest rates of skin cancer in the world and is situated beneath the outer edges of the ozone hole when it is at its maximum in each Antarctic spring. But this apparent correlation may arise simply because the country's largely Caucasian population is ill-suited to the sunny climate and enjoys spending too much time on the beach. Meanwhile, the advice is simple: Put on plenty of sunscreen—the apparently cynical original retort of CFC manufacturers to claims that their product was the problem—and avoid, if you can, the temptation to travel to Antarctica in the spring.

The Short, Hot Summer of 2006

"What hurricane season?"
—*Fox News*

July 2006 was glorious throughout much of Europe. For the UK, the Netherlands, Belgium, and Germany, it proved to be not only the hottest and sunniest July but the hottest month since records began. Toward the end of the month, however, the novelty of the delightful weather was wearing off. On the 27th, having chronicled the heat wave relentlessly over the previous weeks, the *Daily Express* chose this headline for its front page: "August is going to be hotter than July." On a day when other newspapers reported the news—Tony Blair being called to account over the Iraq war, intensified conflict following Israel's invasion of Lebanon—the *Express* saw fit to predict the weather in 120-point type.

"Britain's record-breaking heat wave is set to sizzle on with forecasters predicting an August of 'extreme heat,'" the story began. Reading on, though, it seemed that the forecasters themselves weren't so sure. "The hottest day ever recorded happened in early August and I wouldn't rule out it happening again," said one cautiously.

The headline was remarkable for two reasons. First, it showed how the British obsession with the weather could lead a national daily to forget itself so completely as to give up on reporting actual events in favor of entirely speculative prognostication. Second, August wasn't hotter than July. Far from it. It was one of the most miserable British Augusts in quite a while. (Weather statistics record the recordable—sunshine hours, rainfall, temperature—which does not always reveal subjective things like miserableness.)

But the *Express* was not to be deterred by mere facts. On September 19th, there was once again so little going on in the world that the weather got the front page: "82°F Britain's glorious Indian summer . . . but gales on way." The gales never came. On November 2, after the first autumn frost, the paper tried another tack. A large blue box on the front page had the legend "−14°C" inscribed within it. (It helps to put warm temperatures in Fahrenheit and cool ones in Celsius if you want to make them look extreme. Negative 14°C actually equates to around 7°F—not really all that cold.) Underneath the number was the warning:"Big freeze will hit Britain this month."

Well, did it? Did it hell.Temperatures for November were 1–2°C (2–4°F) above average, rounding off what turned out to be Britain's warmest autumn on record.

Toward the end of the month, the *Daily Mail* observed that the weather had gone "balmy and barmy."The recorded data certainly attested to the first claim. And there was no shortage of bizarre observations to support the second, from flowering snowdrops to horses flying through the air, lifted aloft by a mini tornado.These weather phenomena are noteworthy, even newsworthy for some. But were they freakish?

The most freakish-seeming event of the odd summer of 2006 was that September was warmer than August.This appears odd because it is not simply an extreme but an anomaly, an upset to the natural order of things. Surely it can never have happened before that a summer month was trumped for temperature by an autumn one. Central England temperature records go back to 1659. September 2006 was warmer than the previous record holder, 1729, but there have been Septembers warmer than the preceding Augusts eight times before. It has happened on average every 40 years or so—not that rare, after all.

So what is freak weather? How often should we expect it? And is it becoming more frequent? Surprisingly, perhaps, weather agencies are reluctant to offer definitions.What determines a freak event depends entirely on what's chosen as the normal range. Freak events therefore occur at a constant average rate, which is governed by the defined level of freakishness.When planning protective measures against floods and hurricane, governments find it convenient to talk in terms of ten-year events, or hundred-year events, for example.

However, this model makes a number of possibly unrealistic assumptions. One is that we are dealing with a system in which, although individual events depart from the average, the average itself remains constant. Another is that this average is accurately known in the first place because we have a long record of data. Difficulties and misunderstanding arise if the record period is too short to be representative or if the underlying norms are changing. Public and media attitudes to such events, and toward policy designed to prepare for them and deal with them, may be colored if this is felt to be the case. A former sense of fatalism will be replaced by outrage and demands to know "why weren't we told?" and "why wasn't more done?"

Thus, if global warming is a systematic trend influencing our weather, then we would expect that certain events—heat waves, hot summers, hot autumns, hot years, also droughts and floods—should begin to occur more often. Extremes may become greater, which is to say that the frequency of freak events may increase if our definition of what is a freak stays the same.

It was not only Europe that saw exceptional weather in 2006. China experienced its worst typhoon for 50 years, and there were unusually frequent and severe tornadoes in Kansas. New York City and Japan were blanketed under the deepest snows on record, but Canada had its warmest winter. Phoenix, Arizona, and much of the southwestern United States experienced severe drought. The worst floods for 50 years brought an abrupt end to drought in the Horn of Africa. Sydney had its hottest-ever on New Year's Day.

And this is just one year. Yet according to the climatologist John Houghton, writing a few years ago about global warming, the 1980s and 1990s were also remarkable for the frequency and intensity of their extremes of weather. In 1987, for example, northern Europe experienced its strongest storm since 1703. The following year Bangladesh suffered its worst-ever floods. But, he adds, "a note of caution must be sounded. The range of normal natural climate variation is large. Climate extremes are nothing new. Climate records are continually being broken."[1] Lucky for the newspapers!

Houghton is at pains to distinguish between individual weather events and the underlying climate trend. It is possible to show very simply that these may not be as closely coupled as we are led to think by a media increasingly prone to blame every meteorological oddity on global warming. ("Summer heat waves may get much worse," warned the *Independent*, citing climate change as the cause in September 2006. Six months earlier, toward the end of the winter rather than the summer, the same paper announced "UK winter storms 'to get stronger,'" also due to climate change.)

Assuming for the moment that weather is a very simple system (which of course we know it is not), we can use statistical methods alone to predict temperature extremes against a background of steadily rising temperature. Say we expect a temperature rise of 5.5°F over a hundred years (the average estimated by the Intergovernmental Panel on Climate Change in 2001). How much more often should we then expect a heat wave such as the one that much of Britain enjoyed in July 2006, when the temperature was up by 6.5°F on the average for the month? The answer is that we can expect to see this temperature reached or surpassed about 1 summer in 3—not every summer without fail.

The European heat wave of 2006 prompted headlines, but also health warnings, something largely new in the media treatment of summer weather. These came in response to high temperatures three summers earlier. In 2003, 20,000 people are estimated to have died from heat-related causes in Italy, another 15,000 in France. The death rate in Paris was 8 times the normal, many of those who died being neglected elderly relatives of families who had abandoned the city for their summer vacation. It was said to have been Europe's hottest summer in 500 years. But this does not necessarily make it a freak. Average temperatures have risen since the mini ice age of the seventeenth century, so summers that seemed freakish then might not compete with very hot summers today.

The fact that many people died as a direct consequence of the heat is nevertheless a serious matter. Calling it a freak event does them disservice as it appears to excuse us from responsibility for taking precautionary measures against a predictable and predicted event.

The mismatch between the probability of an event and our preparedness for it can reach absurd levels. How can it be, for example, that Britain in 2006 could suffer "one of its worst droughts for 100 years after two years of below-average rainfall?" Is this remarkable at all? Is it fair enough that we have been caught unawares? Is it cause for outrage? We should expect (at least) 2 consecutive years of below-average rainfall 25 times in a century (on average). So there was nothing odd about this event, and the headlines proclaiming a drought were more a reflection of the increased demand for water due to population migration to drier parts of the country than of any genuine freak of the weather.

This mismatch can also have tragic consequences. Katrina was "only" a category-3 hurricane when it struck New Orleans on August 29, 2005. Category-3 hurricanes make landfall in the United States every year or two. This one commanded attention first in the Gulf of Mexico, intensifying from category-4 to -5 as it turned north before weakening again only 100 miles off the American coast. (Only 3 category-5 hurricanes have hit the U.S. since records began in 1851.) Katrina killed more than 1,800 people and caused damage estimated at over $80 billion—making it America's most expensive hurricane and worst natural disaster—because the scale and severity of the hurricane at sea produced an exceptional storm surge, up to 30 feet at some points along the coast, which breached the poorly maintained levees of New Orleans.

But was it just a strong hurricane that took an unfortunate route, or is it a harbinger of worse to come? Prior to the European summer heat wave of 2003 and Hurricane Katrina in 2005, the environment campaigner Jonathon Porritt could write that "it would be extremely difficult to demonstrate that anyone had yet died from changes in the climate specifically brought about through pollution from our industrial economies."[2] But these extreme events have led some to stop equivocating. For the science writer Fred Pearce, the heat wave was "the first single weather event that climate scientists felt prepared to say was directly attributable to man-made climate change."[3] The evidence for Katrina is more ambiguous because it is the surface temperature of the sea that generates the hurricane. There had been spells of strong hurricanes long before other climate change effects began to be noticed.

Given its profusion of wild weather, we might expect that 2006 overtook 2005 in Atlantic hurricanes as well, as it had been predicted to do. Yet by the official close of the hurricane season on November 30, not a single hurricane had made landfall in the U.S. This non-event is not rare either—the last year to draw a blank was 2001. Much of the media eagerly marked the day. "What hurricane season?" Fox News demanded to know. "Storm predictions prove all wet as '06 season ends," exulted the *Orlando Sentinel*.

But others simply couldn't believe the good luck. The *Miami Herald* was so disaster-addicted as to keep faith with the discredited merchants of doom, warning, "Forecasters predict 7 hurricanes for 2007." A more sophisticated version of this view came from Kerry Emanuel, a meteorologist at the Massachusetts Institute of Technology. "We may see some quiet years—this year may be quiet," he said, before the 2006 season had got underway. But he added, "We probably won't see a quiet decade again in the Atlantic."[4]

Some scientists, including Emanuel, cautiously attribute the low Atlantic hurricane incidence in 2006 to the El Niño event, a warming of the eastern Pacific Ocean that affects winds in the Atlantic. (Others counter that similar conditions were present in hurricane-rich 2004.) The El Niño hypothesis was seized upon by the media, perhaps because the superstitiously named phenomenon hinted at the age-old tendency to attribute extreme natural events, or our deliverance from them, to supernatural causes. "Thank El Niño for mild Atlantic hurricane season," entreated the Reuters news service.

In short, hurricane frequency shows no change from the long-term average. Some studies suggest that the strongest storms, categories four and five, are increasing at the expense of weaker ones, which is consistent with the expectation that climate change will produce more severe weather events of many kinds. But the data are inadequate to be able to say for sure, and scientific opinion remains divided.

Scientists expect that a major effect of climate change will be an increase in freak weather events, from the locally talk-worthy to major disaster-makers. But this does not mean that all freak weather—or even most freak weather events—can be blamed on climate change. For many, though, it remains a strong temptation to make the link anyway, as a lobby group of scientists did in the 2004 United States presidential election campaign in Florida. Battered already that season by Charley, Frances, Ivan, and Jeanne, voters were then greeted by billboards with the slogan "Global warming = worse hurricanes. George Bush just doesn't get it."[5]

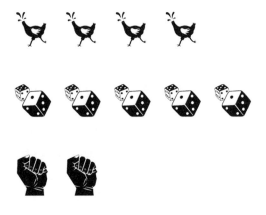

Becoming Unsettled

"A 4th climate warning. Anyone listening?"
—*The New York Times*

Climate change is arguably the most serious issue we confront in this book. It may not have a sudden or immediate impact but it does threaten to affect us all, which has made it the media's most sustained, enduring, and infectious panic story. In the U.S., as a journalism accountability magazine called *FAIR* reports, the media tends to exaggerate the debatability of the issue by presenting both sides of the debate (even though there are far fewer global warming skeptics than global warming believers) and continuing to use scare quotes around the phrase "global warming"—even long after it has become an accepted term. But scare quotes or none, newspaper usage of the phrase "global warming" has been increasing by a steady annual 60 percent in recent years. It was mentioned 16,755 times in the British press alone during June–August 2006.[1]

"Be worried. Be very worried," cautioned the cover of *Time*, showing a polar bear peering over the edge of a very small ice floe. Even the doughty *Economist* decided that "the heat is on." Both magazines dedicated special surveys to the topic in 2006. The flow of alarmist headlines has become so copious that one think tank was prompted to lash out against "climate porn."[2]

How much should we worry? Is climate change a greater threat than terrorism, as scientist David King has insisted in his effort to ram home the danger? Almost certainly, it is—and not only because the risk of being killed in a terrorist attack is actually very small, as we saw in Chapter 6. But threat of what? With terrorism the nature of the hazard is clear, but with climate change we don't yet know quite what we face.

There is, however, no shortage of respectable oracles to warn that the worst will happen. The wrist-slitting testament of James Lovelock, *The Revenge of Gaia*, even Friends of the Earth found "gloomy." For Lovelock, "the fever of global heating is real and deadly and might already have moved outside our and the Earth's control." It is too late for sustainable development—"what we need is a sustainable retreat."[3] Al Gore's film *An Inconvenient Truth* is hardly less bleak. "I look around for meaningful signs we are about to change," he says forlornly. "I don't see it right now."

According to ever more extreme forecasts, temperatures may rise by up to 9, 14, 20 . . . degrees Fahrenheit over the next hundred years, disrupting agriculture, plant and animal life, and our vacation plans. In Britain, people don't know whether to laugh or cry: "The [Mediterranean] may get too hot for holidays," but, then again,

"Norfolk to be the new [Mediterranean]."[4] But in already arid regions of the world the effects would be dire. The rising temperatures will melt the polar icecaps, raising sea levels by half a foot, 3 feet, 20 feet, 200 feet . . . The heat will push climate systems into overdrive, producing longer, drier droughts, heavier downpours, more sudden floods, more vicious storms . . . No wonder there's now a "coalition of environment, development, faith-based, women's, and other organizations" piteously terming itself "Stop Climate Chaos."

The cause of all this is believed to be the increasing concentration of greenhouse gases in the atmosphere, most notably carbon dioxide (CO_2) produced by humankind's burning of fossil fuels. Believed, that is, by virtually all scientists—and disbelieved by a small minority who hold either that there is no trend of rising temperatures, or that if there is it is due to terrestrial natural causes or else solar effects.[5] However, the balance of likelihoods now leans very strongly toward man-made carbon dioxide, as we shall see.

In a trenchant 1995 skeptics' handbook, *But Is It True?*, Aaron Wildavsky quoted a 1981 paper in the journal *Science* by James Hansen of NASA's Goddard Institute of Space Studies in New York predicting that "potential effects on climate in the twenty-first century include the creation of drought-prone regions in North America and central Asia as part of a shifting of climatic zones, erosion of the West Antarctic ice sheet with a consequent worldwide rise in sea level, and opening of the fabled Northwest Passage." Hansen added that evidence of warming due specifically to man-made carbon dioxide "should emerge from the noise level of natural climate variability by the end of the [twentieth] century."[6] Wildavsky's purpose in 1995 was to mock the ominous nature of these visions. Yet in 2007 they were all happening. As late as 2000, CNN reportage was just beginning to put together the pieces: In an article called "Arctic Expert Unthaws Alarming Data on Thinning Ice," one reporter "speculates" as to whether such data is an indication of global warming. And yet skepticism sometimes prevails:

In 2003, the *Sunday Telegraph* gloated that climatologists' disastrous predictions for the Arctic icecap were falling "foul of reality." "The latest research points to a deepening of the polar freeze," the article assured the public. Since then, melting has accelerated faster than even many scientists were predicting.

It's a good instinct to doubt, and a better one to test for yourself. Try checking the temperature records where you live. Check the national records. Or maybe you don't think that temperatures measured here and there are a reliable indicator of the heat received by the Earth. Then ask gardeners and birdwatchers whether they've noticed any changes.

Ask yourself about the 2,000 or more scientists who contribute to the United Nations Intergovernmental Panel on Climate Change, the body that makes the most influential predictions in the field. Is it just a "quango-cum-travel-agency for those whose salaries depend on keeping the world worried about global warming," as some columnists think? The vast majority work for universities or national institutions dependent on government funding. Half of them owe their salaries to the American taxpayer. How can the IPCC be a conspiracy when its findings so clearly do not suit the governments that support it? Wouldn't governments arrange matters so as to hear better news?

From chemical analysis of rock samples and ice cores which contain historically trapped gas, scientists believe that the concentration of carbon dioxide in the Earth's atmosphere was rather stable at around 280 parts per million (ppm) for more than 20 million years. In 1958, Charles Keeling of San Diego's Scripps Institution of Oceanography began taking direct readings at the summit of Mauna Loa in Hawaii, far from any source of pollution. Allowing for the carbon dioxide taken up by oceans and vegetation, each year showed a small, steady rise by an amount corresponding closely with the quantity of carbon-based fuel burned around the globe. When Keeling died in 2005, his employer called his figures "the single most important environmental data set taken in the 20th century."[7]

Carbon dioxide in the atmosphere was constant until about 1800. The rise began modestly in the nineteenth and early twentieth centuries at a rate of 0.25 ppm a year. When Keeling began his work, it stood at 315 ppm. Now the annual rise is at least 1.5 ppm. When the figure was announced for 2006—383 ppm—the London *Metro*, armed with a quote from David King, made a poorly judged attempt to heighten the alarm: "CO_2 levels 'hit 30m year high.'" But the drama is not in comparison with the state of affairs long before the evolution of the human species, it is in the inexorable increase since the beginnings of industrialization.

None but the most recalcitrant dieselhead would contest these data or the deduction that this carbon dioxide must be man-made. But what about temperatures? The long view was laid out in 1999 by Michael Mann, a statistical climatologist at the University of Virginia. He pooled data gathered by modern instruments with temperatures going back a thousand years "reconstructed" from analyses of tree rings, corals, and ice cores. The resulting graph has become known as the hockey stick because of its sharply upturned right-hand end, representing recent rising temperatures after a long period of stability. It drew fire from climate skeptics for its use of proxy data gathered from disparate sources. But in the absence of anything better, it would surely be remiss not to attempt such a study at all—and besides, economists and politicians routinely use this method. Despite criticism, Mann's graph has been substantially confirmed by more recent studies, and the 2007 report of the IPCC features an improved version of the same basic trend line. This latest report expects temperatures to rise between 1.8 and 11.3°F by 2100, with around 7°F judged most likely.

The hockey stick graph of actual and predicted rising temperatures and that for the concentration of carbon dioxide in the atmosphere look remarkably similar, strongly suggesting (though not proving) a correlation between the two. However, because the upturns in the two graphs are so recent—seen in only the last few decades over a span of ten centuries—there remain uncertainties about the extent to which temperatures will follow the rising carbon dioxide trend and the time delay in their doing so. These have vital implications, not only for understanding the scale of global warming but also for deciding how to deal with it.

Climate studies indicate that if carbon dioxide emissions ceased tomorrow it would still be 10 to 20 years before the last molecules of the gas were effectively distributed through the atmosphere to exert the maximum greenhouse effect. It would be another 30 to 50 years before the infrared radiation they absorbed was passed on via other constituents of the air to produce heating at the Earth's surface—although ice core studies of earlier increases in carbon dioxide suggest that the full temperature effect might not be felt for 1,000 years.

Add to this uncertain wait the greater uncertainties (but potentially greater risks) associated with what have become known as tipping points—phenomena of positive feedback where an effect produced by warming in turn promotes further, perhaps sudden, warming. The list seems ever-growing: As polar ice melts, reflective snow is replaced by dark sea which absorbs more heat; as peat in the soil warms it releases methane, another greenhouse gas; if plants die from overheating, they cease to absorb carbon dioxide or, worse still, release it if they are consumed in forest fires; and so on. If we pass these points, it is claimed, it will be impossible to reverse the damage no matter what we do.

Let us be more positive.

First of all, will things actually get worse? Newspapers always emphasize the negative, so front-page headlines warning, say, of "the century of drought" are to be expected. Potentially increased crop yields or reduced winter deaths in other parts of the world go unremarked. Nevertheless, there are sound reasons to expect that the human situation will become overall more acute, if not as desperate as the media portrays—reasons such as net loss of stable, temperate climate zones, a more energetic climate system powering extreme weather events and, not least, the upheaval of change itself. Smaller environmental changes than those we may be confronting have been sufficient to destroy local civilizations during the past several thousand years of relative climate stability. This time, we face either "quick and possibly costly adaptation" or migration, "which has become difficult or, in some cases, impossible in the modern crowded world."[8]

Despite this, the Earth as a whole remains highly habitable, much of its surface amply adapted to human life. The parts that are best adapted will doubtless shift a little, but climate change over the next century need not be a catastrophe and could prove no more than a moderate inconvenience for most people. Other parts will become less genial, but it's not as if this is a novelty. Enough societies struggle to survive today in areas already maladapted to human life, whether due to natural forces or human intervention.

How do we adapt? The odd bedfellows of the Stop Climate Chaos campaign group reflect the uncertainty: Do we go back to nature, do we try progress, do we just pray? Climate change is an issue that affects us all, and there is an understandable feeling that we should all contribute toward resolving it. But the present focus of personal effort on using low-energy light bulbs and switching off appliances on "standby" seems woefully at odds with the scale of the problem. It may even be unhelpful. James Hansen has warned that without a government hand on the tiller, "conservation of energy by individuals merely reduces the demand for fuel, thus lowering prices and ultimately promoting the wasteful use of energy."[9]

We need to think bigger. Faced with a large problem, it sometimes helps to break it into more manageable chunks. In 2004, Stephen Pacala and Robert Socolow, respectively an ecologist and an engineer at Princeton University, looked at the upward graph of carbon dioxide emissions, projected to rise to 14 billion tons of carbon burned per year by 2054, and the horizontal line, if levels could be held where they are at present, at 7 billion tons a year. They divided the fat wedge between the two lines into seven equal thinner wedges, each representing a reduction of 1 billion tons a year by 2054. They then identified not seven but fifteen existing technologies that could each account for a wedge of carbon.[10] From staring disaster in the face, we were suddenly shown to have the luxury of choice in how to avoid it. The ideas are split between reducing energy demand (halving car miles, doubling car fuel efficiency, improving building insulation, for example), substituting carbon-free energy sources (nuclear, solar, wind), and capturing carbon dioxide (planting trees, sequestering power station emissions). What Pacala's proposal makes clear in its unideological way is that the solution to the problem is unlikely to be either entirely technological or entirely the self-denying ordinances of the dark green environmentalists. It's a mnemonic more than a prescription, but it shows that a start can be made.

How much it will cost to make such wholesale changes is anybody's guess. There will in any case still be an additional cost to adapt to climate change already on the way that mere stabilization of carbon dioxide emissions (as opposed to a cut) does nothing to forestall. And of course, doing nothing also carries an economic cost. However, there is in fact reasonable agreement among estimates of the cost of addressing climate change arrived at by sources ranging from Friends of the Earth to multinational accountancy firms. (There is, as you might expect, greater divergence in estimates of the cost of doing nothing, although this cost is reckoned to be at least as great as, and by some very much more than, the cost of mitigation.)

As climate change becomes a political and economic issue, and as it is increasingly realized that some climate change is inevitable, the phrase "dangerous climate change" has entered the lexicon. But this is a largely meaningless concept. Norfolk's "new [Mediterranean]" is presumably not "dangerous climate change," but then Norfolk's new Mediterranean may be the Mediterranean's new Sahara. It is not sensible therefore to attempt to put a price on preventing dangerous climate change. However, economists do agree on the cost of limiting the global average temperature increase. Capping the rise at 3.6°F will come out at between 1 and 2 percent of global GDP. This puts the price tag in the area of $500 billion to $1 trillion a year, or around 100 dollars, euros, or pounds per person per year—hardly unmanageable, especially if the load is distributed more equitably toward the rich countries responsible for the emissions.

Pigs Might Swim

"American shores face threat of rising sea level"
—*ABC News*

With its white beaches, scattered palm trees, and lightly clad inhabitants, the tiny country of Tuvalu in the South Pacific Ocean looks in newspaper photographs like an island paradise. But there is trouble in paradise. The eight little atolls that make up Tuvalu rise only 16 feet above sea level and are in the front line of global climate change. While their politicians tour the world lobbying unsuccessfully for cuts in carbon emissions, Tuvaluans wait nervously for the next high tide that will wash salt water over their crops and round the trotters of their grazing pigs.

"Tuvalu envoy takes up global warming fight," intoned NPR's Morning Edition. More controversially, one recent British newspaper headline read, "Sinking islanders are facing mass evacuation," although of course neither the islanders nor their islands are actually sinking. It is the sea that is rising. According to Mark Lynas's colorful *tour d'horizon* of parts of the world already feeling the effects of climate change, *High Tide*, Tuvaluans now face the unenviable choice of whether to move and live "cultureless and uprooted in a foreign country, or stay on the land of their forefathers and die."[1]

Sea levels have risen by about 8 inches over the last century. They are rising on average (the effect is not evenly spread around the world's oceans) by 1 to 2 inches a year at the moment. The 2001 report of the UN's Intergovernmental Panel on Climate Change (IPCC) predicted that sea levels might rise by 19 inches by 2100.[2]

There are a number of sources of sea-level rise, although the relative contributions of each to the overall effect are still the subject of scientific debate. Probably the greatest contribution, though the one likely to be forgotten by most of us, comes from the thermal expansion of the oceans due to global warming. This is thought to account for about 60 percent of the currently observed rise. Then there is the more theatrical effect of melting Arctic and Antarctic ice. Melting ice floes do not alter the sea level as they already displace water, but the ice sheets and glaciers on land that melt and run off into the sea make up the remaining 40 percent. However, these ice sheets and glaciers are thawing faster than was once predicted, and over the next hundred years melting of the Greenland icecap alone could add at least as much again. As white ice melts into blue sea, the Earth's surface becomes that much darker, and so absorbs more sunlight, producing a greater global warming. In some scientists' view the reduced reflectivity of the Earth is now beginning to contribute more to global warming than our injection of carbon dioxide into the atmosphere.

A negligible additional factor in rising sea levels is due to melting mountain glaciers in temperate and tropical regions, such as the Himalayas, Andes, and the fabled snows of Kilimanjaro, which some have predicted will disappear within as little as ten years. These add 25 cubic miles a year to the total, which may be "more than the entire volume of Lake Geneva," but is still almost literally a drop in the ocean in terms of sea-level rise.[3] This melting is more significant for the fact that it jeopardizes the water supply upon which many mountainside communities depend. As much as 40 percent of the world's population depends ultimately on Himalayan melt-water for its drinking supply, for example.

These figures seem modest: 19 inches is about up to the knees, and it'll take 100 years, a slow rising tide by any measure. Certainly these figures are not scary enough for the media. *Time* magazine, in a special report on global warming in April 2006, explained these causes of sea-level rise, but the figure it chose to quote in its text for any actual rise was the 23 feet due to the melting of the entire Greenland ice sheet, "swallowing up coastal Florida and most of Bangladesh." (Elsewhere, a graph presented the accepted IPCC projections without interpretation.)

Fred Pearce, the veteran environment writer for *New Scientist*, chose the same lurid figure for an article in the magazine in 2000 based on the then forthcoming 2001 report of the IPCC. The piece ran under the preposterous headline "Washed off the map: Better get that ark ready because sea levels are gonna rise." Al Gore's film, *An Inconvenient Truth*, spared us the ark but also could not resist the 23-foot deluge.

When the content of the story fails to provide drama, the art department is often happy to help. For instance, the *Daily Telegraph* in December 2006 reported the view of Stefan Rahmstorf of the Potsdam Institute for Climate Impact Research that sea levels might rise by 3 feet more than is generally expected over the next century. The article was generously illustrated with montages of sodden London landmarks. One showed water lapping gently against the masts of the Millennium Dome.

Even a 23-foot rise in sea level will not happen for quite some hundreds of years under the most pessimistic scenario. But this has not stopped even level-headed scientists from joining the more excitable media and showing us what the world would look like in this eventuality—a simple matter of plugging the relevant topographical data into a computer graphics package. The results—typically showing the coasts of Florida or East Anglia or the Netherlands—impose sea levels projected for hundreds of years hence on top of present-day cities and take no account of protective measures that would surely have been implemented during the passing centuries in these wealthy regions.

The morbidly curious might like to know at this point that if *all* the ice in the world were to melt it would raise the ocean level by 230 feet—a calamity that requires temperatures far in excess of even the most apocalyptic forecasts and even then would take thousands of years to transpire. Although there is not the remotest prospect of this happening, this too is a factoid routinely thrown in to juice up stories on rising sea levels and featured in the recent special reports of *Time* and the *Economist*.

Yet it's not as if the more modest rise is without consequence. A sea-level rise of no more than three feet would displace up to 100 million people in Bangladesh, Nigeria, China, and Egypt alone and inundate much productive agricultural land.[4] Interestingly, Florida and the Netherlands, with large populations living at or below sea level already, tend to be omitted from such round-ups, presumably because it is taken as read that they will be defended, for yes, prosperity protects even against the rising sea.

Sea-level rises are clearly felt most at the highest tides. It was a combination of high tides with unfavorable winds and atmospheric pressure that led to the disastrous coastal flooding in the Netherlands and East Anglia in 1953 and in Venice in 1966. This is a great help to the media, as news editors need equip themselves with nothing more than a tide table to work out when to run the story. The headlines can be used over and over. "Highest tides in 20 years threaten coast towns this weekend" appeared in September 2006 in the *Times*, accompanied by a photograph of a man keeping "an eye on sea levels" with the aid of a pair of binoculars.

Sea-level rise attributable to global warming is exacerbated in some populous parts of the world by the fact that the land is moving independently. Along the eastern seaboard of the United States and in southeast England, for example, the land is sinking at rates similar to those at which the sea is rising as land further north, once depressed by the weight of ice-age glaciers, seesaws upward.

In such places bulldozers have been used to bank up protective shingle barriers. But these may be of little use. Astonishingly, each thirty-second-of-an-inch rise in sea level may push these barriers back by as much as 3 feet due to the complex workings of the processes of erosion.[5] On exposed coasts, this means that the shoreline can be expected to retreat by a pictorially gratifying 10 to 13 feet a year. Add to this the prospect of higher waves and stormier storms as the side effects of global warming, and a fraction of an inch more of a calm sea has been transformed into a vindictive assault on our coasts.

It was King Canute's subjects who demanded that he command the sea to stop rising. The king politely turned up on the beach and, having already said there was nothing he could do, proceeded to demonstrate his powerlessness before the tide. Today, people in coastal communities appeal not to monarchs but to governments. Having for decades adopted the position that existing coastlines are to be defended, however, governments are increasingly shifting toward a policy of strategic retreat or, more euphemistically, "managed realignment." Major settlements will continue to be protected, but in some places the sea is being admitted deliberately by breaching long-standing barriers. The aim is to sacrifice a coastal band of agricultural land to create a wide barrier of marsh that will offer a degree of natural flood protection. In some cases, the policy is defended not on economic grounds but because continuing to maintain the sea defense would cause wildlife habitat to be lost—cold comfort to the humans whose own habitat may be sacrificed in the process.

But for most of us, rising sea levels are the least of our worries among the effects of climate change. They affect relatively few people and are happening so slowly it's almost invisible, certainly far more slowly than the communities concerned are able to react. "Sea level rise is such a slow process that once started it's almost impossible to reverse," warns Mark Lynas in *High Tide*.[6] But his absurd nonsequitur gives the game away: There is nothing intrinsically unstoppable about something slow, indeed there is more time to stop it. Nevertheless, inexorability is an essential journalistic makeshift when things aren't happening with catastrophic speed. In March 2006, the *Baltimore Sun*—sea levels are rising comparatively fast at just over one-tenth of an inch a year in Baltimore as along the rest of the U.S. eastern seaboard—attempted to alarm its readers by reporting the words of a local professor in environmental science: "The seas are expected to rise slowly and steadily, but no one knows how soon or by how much."

Indeed, scientific uncertainty is very high. The IPCC's 2001 report anticipated that sea levels would rise by 19 inches by 2100, but this is the midpoint of a range between 3.5 and 34.5 inches, an order-of-magnitude variation in estimates. This reflects the simple lack of knowledge of all that happens in complex natural systems. It's not clear yet, for example, whether the accelerating melting observed in the Arctic will be replicated in the Antarctic (which contains 90 percent of the world's ice). At the moment it is generally thought that the Antarctic is in fact helping to mitigate rising sea levels by receiving larger net snowfalls each year than it loses due to melting. Nor is it clear how surface and groundwater around the world fit into the picture. These uncertainties are why the IPCC's 2001 predicted rise of 19 inches was in fact a reduction on the prediction of 21 inches in the IPCC's previous report of 1996. The estimate was cut again, to 17 inches, in the IPCC's most recent report in 2007, although some reputable studies published since the cutoff date for this report favor a significantly increased estimate of around 31.5 inches.[7]

We may not have taken the 8-inch rise during the twentieth century quite in our stride. It took the deaths of some 1,835 people in the Netherlands and 307 in Britain to prompt the launch of the Dutch Delta Works scheme of national flood defense and the construction of the Thames Barrier. But at least these so far adequate defensive measures were taken; the Venetians have yet to act to protect the Serenissima. This work will surely continue where it is feasible.

Meanwhile, populations that face more serious disruption are showing a good sense that evades the newspapers. The Tuvaluans have brokered a deal with New Zealand to resettle the country's entire population. Other, safer, South Pacific islands have also offered refuge. Are the islanders scrambling to escape? Hardly. New Zealand offered a quota of 300 immigrants a year, which would see the entire population of 11,000 resettled in little more than a generation. The Tuvaluans have suggested a lower rate of 75 people per year so that essential social services may be maintained on the islands. The Tuvaluans are in effect hedging their bets, which seems only wise given the high uncertainties in calculating future sea levels.

The Maldives are even lower than Tuvalu and home to more than 300,000 souls. Here the strategy is to concentrate areas of population into defensible enclaves. "In Bangladesh the future has arrived," according to the country's High Commissioner, Sabihuddin Ahmed.[8] "Climate change will eventually threaten thirty to forty million lives [a quarter of the population] in Bangladesh. When these people's homes and crops are flooded forever, where will they go?" Where they go is indeed a serious issue, but there is time to plan. The timeframe over which sea levels are predicted to rise dangerously—in the Thames Gateway, the Netherlands polders, the Florida coast, and even in Bangladesh—is far greater than the duration of the typical housing policy or even the lifetime of some modern dwellings. The Netherlands needs no new technology to guard against sea-level rise. Scientists have estimated that spending $12 billion on raised earthworks and greater pumping over the century would be enough to deal with a 3-foot rise—that's about $15 a year for each Dutch citizen. If Bangladeshis grow wealthier at the predicted global rate, they too will be able to afford to safeguard many homes, greatly reducing the numbers of those who will need to migrate by the time the forecasts become a reality.

Go with the Flow

"Risk of quakes adds spice to life"
—*San Francisco Chronicle*

Around the world, hundreds of millions of people live with the daily risk of extermination by the Earth's geophysical might—500 million of them from volcanic eruption, 130 million as the result of earthquakes, according to separate estimates (another—clearly incompatible—estimate puts 75 million Americans in 39 states at significant risk from earthquakes).[1] Many millions more are within realistic striking distance of the tsunami that would follow a massive geological upheaval under the sea. These numbers are growing faster than the rate of increase in global population. Every year, up to 80 million people are added to the numbers at risk from some kind of natural disaster, according to the United Nations' International Strategy for Disaster Reduction. It seems that we are choosing to live dangerously.

Even in these modern times, our violent planet has us in its thrall. Earthquakes and especially volcanos, with their visual sublimity, are two of the staple scenarios of the disaster movie. No hard sell is needed in order to inspire terror beyond the bare words—*Earthquake* came out in 1974, and *Volcano* in 1997. *Tsunami* followed in 2006.

For some, though, the cinema isn't close enough. Volcano tourism is a growing business. Mauna Kea on the Big Island of Hawaii competes with Vanuatu, with holiday brochures advertising the world's safest volcano. The volcanoes of Italy—Vesuvio, Etna, Stromboli—became essential stops for artists and writers on the Grand Tour. Today, thousands of tourists climb Etna's slopes each year. Injuries and deaths are a regular occurrence. Nine people were killed in 1979, and another 2 in 1987, when they wandered too close to active vents, adding markedly to the rather modest total of 55 recorded lives lost due to Etna's volcanic activity in all the preceding 3,500 years.

But these are mere teasers for the main event. Geological surveys indicate that there are about 500 known active volcanoes on land. Between them they are responsible for about 60 eruptions a year. Both the most violent and the most famous eruptions happened in the nineteenth century in Indonesia. The Tambora eruption of 1815 was the most powerful ever recorded, sending ash and gases into the atmosphere such that 1816 became known on the other side of the world as the year without a summer. Krakatoa famously destroyed an entire island in 1883. To these 500 may be added volcanoes under the sea about which much less is known, and the knowledge that volcanoes officially listed as dormant can explode back into life. The greatest likelihood of a human disaster today may come from a volcano that everybody considers to be harmless. Vesuvio was just such a volcano until its eruption in CE 79 took the lives of more than 3,000 inhabitants of Pompeii and Herculaneum.

The volcano has scarcely slept since. A recent report revealed archaeologists' discoveries concerning not the eruption of CE 79 documented so colorfully by Pliny the Younger, but an even more powerful eruption known as the Avellino catastrophe which happened 2,000 years earlier. This eruption destroyed the Bronze Age settlement of Avellino but killed relatively few people. Paths found in the ash suggest that most managed to walk to safety.

Today, some 3 million people inhabit the area devastated by this eruption, which includes Naples, Pompeii, Ercolano, and more than a dozen other towns. A fifth of them live in the "red zone" on the lower slopes of the volcano, many in illegally built homes in the Vesuvio National Park. There might be enough warning of an eruption to allow for an orderly evacuation, but a sudden explosion would leave no time for all these people to escape, especially if—as was the case in 1983 at nearby Pozzuoli when residents feared that a small earthquake presaged a major eruption—they try to get away by car and get stuck in the traffic.

Mauna Loa on Hawaii, the world's largest volcano, erupts on average every 10 years or so, but has not done so since 1984, leading to speculation that a big eruption is overdue, especially when seismic activity increased in 2002 and 2004. The major town of Hilo, the tourist coast, and new housing on the volcano's slopes are all within four hours, and in some places just minutes, of being overwhelmed by lava. There is only one road out of the danger area, and, if that were cut off, there would be no means of evacuation other than by helicopter, which would be dangerous and slow. Nevertheless, land is cheap and building is legal, although discouraged by measures such as the withholding by the state government of utility services. The United States Geological Survey estimates that there has been more than $2 billion worth of construction since the eruption in 1984.

The mentality is the same in earthquake zones. "Risk of quakes adds spice to life," headlined the *San Francisco Chronicle*, writing on the centenary of the earthquake and fire that claimed 3,000 lives in the city in 1906. Given a relatively gentle reminder of the ever-present danger in 1989 ($10 billion of damage, 63 killed), San Francisco has invested heavily in its infrastructure, but there remains a severe risk especially to housing from even a moderate quake. Less than 1 person in 7 has earthquake insurance, according to the *Financial Times*. "Most owners are simply trying to ignore the danger." As at Mauna Loa, government interference—banning building or making insurance mandatory—would be seen as an infringement of liberty.

Reading this in New York or Paris or Berlin, you might puzzle at these people's fatalism while quietly congratulating yourself on living somewhere sensible. But you don't have to live along a major fault line for the press to give you the quivers. "London could face quake of LA scale, say geologists," the *Sunday Times* reported in 1996. "Britain 'due a big earthquake,'" the *Financial Times* echoed a few years later. But "big" turns out to be relative—in the geological backwater of the British Isles it means a quake like the one in 1580 in which, it was said, "two people were killed by falling masonry."

Londoners do not really fear earthquakes as they go about their daily business, but then nor do San Franciscans or Angelenos. Why not? For many, the lifestyle benefits outweigh the risk. For others, economic necessity forces the gamble with nature. Plots of building land on the barren slopes of Mauna Loa have been advertised on the Internet for as little as $400. Between major events, people believe that earthquakes and volcanoes are less of a danger than they are in fact, and a bargain is hard to resist.

What changes when the big one strikes? Does a rational reappraisal of the danger take the place of ignorance and denial? Is fatalism banished—or strengthened?

In 1995, the first known eruption of the Soufrière Hills volcano on the Caribbean island of Montserrat destroyed half the island, including its capital, Plymouth, and killed 19 people. A decade later, more than $300 million has been spent redeveloping the island, but few of the 8,000 who left after the eruption—two-thirds of the population—have returned. A survey for the British Department for International Development found that, although most of the Montserratians who fled to Britain "might return home, hardly any are making active plans to do so."[2] It is less the risk of another volcanic eruption that puts them off than the uncertainty of finding a home and a job.

The 1995 earthquake in Kobe claimed more than 6,000 lives and did $250 billion worth of damage, but it also awakened the Japanese to the truth that their buildings were not as earthquake-proof as they had been led to believe. How to prevent a recurrence was made obvious by studying the pattern of the damage. As is generally true in earthquakes, most of those who died were crushed by collapsing buildings. It was duly found that nearly all of these buildings had been put up before 1981, when building regulations were toughened. Rebuilt according to the new codes, Kobe at least should be far better equipped to withstand another earthquake of the same intensity.

But what of the capital, Tokyo, which, according to the *Financial Times*, "is overdue a potentially enormous earthquake?" Since Kobe, the Japanese have greatly increased the number of seismic monitors to provide better warnings. But, according to Bill McGuire, a professor of geophysical hazards, "Nobody has ever accurately predicted an earthquake there."[3] Although new buildings conform to the latest standards, older ones do not. McGuire estimates the cost of a quake may be more than $3 trillion to rebuild and 60,000 dead—grim, but less than half the toll of the last major Tokyo quake in 1923. Tokyo city authorities have also improved procedures for managing a disaster, but there is nevertheless a recognition, reinforced by experience at Kobe, where it was not rescue teams but simply people at the scene who saved most lives, that people "ultimately have to be responsible for themselves."[4]

Planning for catastrophic events that might happen at any time is not straightforward. A general warning may have no effect, while a specific warning can only be given on the strength of a firm prediction, which is often impossible to make. The massive eruption of Mount Pinatubo in the Philippines in 1991 was predicted and an evacuation successfully accomplished, saving thousands of lives. It is not insignificant that there was an American air force base in the affected zone.

When the Colombian Nevado del Ruiz erupted in 1985, by contrast, 25,000 died because the local agencies failed to act despite informed predictions in the month preceding the event. Those living nearby had not associated the volcano with any danger partly because it is hidden from view behind other mountains. No evacuation was organized perhaps for fear of a backlash if no eruption came. The Ecuadorian volcano Tungurahua is one of the world's most active at the time of writing. An eruption in August 2006 destroyed villages and left 1 person dead and 60 missing. In this case, the area had been evacuated the previous month, but people had returned to their homes.

The Naples area authorities are trying to persuade red-zone residents to move out with inducements of €30,000 per household (almost $40,000 in U.S. currency), rather than depend on an evacuation plan that would struggle to cope with a mass exodus. But if the experience at Pozzuoli is anything to go by, people might calculate that it is better to stay put and then demand the government build them a new town after the disaster. The regional president, Antonio Bassolino, wants the residents to go, but at the same time he would like tourists to come. It is even planned to convert some of the evacuated homes into hotels. "Tourists obviously will leave at the volcano's first cough," he says.[5] But volcanologists say he is reckoning without the likely influx of volcano rubberneckers. Neapolitan lives may be saved, but if the hotels fill up perhaps not many lives overall!

The Vesuvius eruption that destroyed Pompeii ranks a mere 5 on the volcanic explosivity index (VEI) analogous to the Richter scale. Like Richter's, this scale is logarithmic for intensity and also for the likelihood of an eruption. Thus, a 6 such as Mount Pinatubo or Krakatoa has 10 times the power of a 5, but whereas a 5 can be expected every decade or so, a 6 is a once-a-century event. We can expect a 7 eruption every millennium, an 8 every 10,000 years, and so on. These eruptions throw out so much ash and gas that, like Tambora in 1815, they can affect the entire planet. Even the Pinatubo eruption managed to depress the global temperature by 1°F for several years. The last eruption with a VEI of eight was more than 20,000 years ago. One need hardly add that "The volcano that could wipe out life on Earth," in the *Daily Mail*'s headline, is "due to erupt any day now."

Chilling News

"Whatever happened to global cooling?"
—*Discover*

With global temperatures on the rise by many measures, ice age warnings have been out of fashion lately. But this was not always the case. In 1973, the magazine *Science Digest* wrote: "The world's climatologists are agreed . . . that we do not have the comfortable distance of tens of thousands of years to prepare for the next ice age . . . once the freeze starts, it will be too late." A couple of years later, a *New York Times* headline ran thus: "Scientists ponder why the world's climate is changing: major cooling widely considered to be inevitable."[1] Popular science books and television documentaries amplified the alarm. The most notorious of these, Lowell Ponte's *The Cooling*, claimed that "cooling has already killed hundreds of thousands of people in poor nations . . . If it continues, and no strong measures are taken to deal with it, the cooling will cause world famine, world chaos, and probably world war, and this could all come by the year 2000."[2]

What would bring on a new ice age? It depended on who you read, but there was no shortage of choices. Climatologist Stephen Schneider cited the possibility of an extreme snow deluge or disintegration of Antarctic ice sheets. Science writer John Gribbin's *Future Weather* discussed possible fluctuations in the output of the sun and dust in the atmosphere. The astronomer Fred Hoyle thought meteorites might do it.

All were writing during a cool spell in the world's climate. The 1910s to the 1940s had been unusually warm, creating dustbowl conditions in the American Midwest. But thereafter, it was cooler than average long enough for climatologists to "become used to the idea that the world was in a cooling phase," as Gribbin admits in a later, rather different book on climate change, *Hothouse Earth.*[3]

An ice age is perhaps the climatic change we can imagine most clearly, partly because we have evidence of what previous ice ages were like, and partly because it brings with it the visible transformation of water, the staff of life, from a usable liquid to an inaccessible solid—a dramatic and disastrous change of state that has no equivalent in global warming or other major climatic shifts. We have a folk memory of the Little Ice Age in the seventeenth and eighteenth centuries, reinforced in the paintings of Bruegel and Avercamp and in stories of the Pilgrim Fathers being helped through their first winter in the New World by the native Americans. And, at least in cultures that experience winter, we possess a darker mythology of the cold personified by ice queens and ice maidens, snowmen and Jack Frost, which has no warm-world counterpart.

Scientific evidence obtained from a range of sources indicates that ice ages occur every 100,000 years or so, punctuated by shorter interglacial periods lasting 10–20,000 years. Put another way, the Earth is a world of ice where, every now and again, circumstances conspire to produce a temporary thaw.

These circumstances relate to the Earth's changing position and orientation relative to the sun. It is only an approximation to say that the Earth orbits at a constant distance from the sun and therefore always receives the same quantity of solar radiation to heat it. In fact, the planet's orbit is slightly eccentric, and the axis about which it is spinning wobbles in various ways. These wobbles and eccentricities exhibit cyclical patterns called Milankovich cycles after the Serbian mathematician who spent his life investigating them. Calculating the periods of these cycles, and therefore how they overlap, is still no simple task, never mind explaining how much each of them affects the sun's ability to heat the Earth's surface, which is one reason why there is such uncertainty over what influences our climate.

In 1982, Gribbin was clear, however. Our interglacial holiday is almost over, and "we are moving rapidly into an orbital configuration appropriate for a full Ice Age." A string of bad winters might be all it takes: "Northern Hemisphere summers are already cool enough for the ice-sheets to remain if once they become established."[4]

There are shorter-term, but still severe, fluctuations in temperature that also challenge science. About 12,000 years ago, toward the end of the last ice age, there was a millennium-long cold spell known as the Dryas (after the tundra flower whose fossil pollen has been used to analyze the phenomenon). And then, 8,200 years ago, came a shorter cold snap. The prevailing theory is that both events were triggered when reservoirs of glacier meltwater gushed into the ocean, although the relevant dates don't correspond perfectly. This ice-cold water, the theory goes, interrupted the so-called ocean conveyor, the circulation of water (and heat) among all the seas, thereby preventing the Gulf Stream from warming the North Atlantic. Although scientists have known about the Dryas event for nearly a century, it is only since 1990 that it has been explained as a side-effect of a broader warming trend. Before that, it was simply viewed as the stuttering last gasp of the ice age.

Revived fears of a new ice age depend heavily on this still poorly understood mechanism thought to be responsible for the Dryas event. This time, fast-melting Arctic ice would flick the Atlantic switch. The popular term "ice age" used in this context is misleading, as the change would not affect the entire northern hemisphere as a full ice age does, nor would it last as long as a full ice age. The world as a whole might get warmer, but northern Europe would cool—and quickly.

Outlandish as they may sound, predictions of a drastic cooling do deserve serious consideration. The conventional method of evaluating risk is to look at the probability of the risk combined with the impact of the threatened event if it does happen. The probability of an abrupt cooling, even in northern Europe, may be low compared to the likelihood of other climate change. But its impact—on agriculture, and on where and how people live—could well be more ruinous than a few degrees of global warming.

Unfortunately, we know little about how the oceans work. We have not had long to observe them, and there is only indirect evidence of their past behavior, for example in ice core samples. Like the atmosphere, the oceans are a complex system in which huge forces are engaged. We know something of the ocean surface from satellite observations, but little about what goes on beneath, where the sheer mass of water, its movements and its stored heat are colossal. For example, just 10 feet of the oceans' depth has the same heat capacity as the entire Earth's atmosphere, yet the depth of ocean that effectively mixes with the atmosphere is 10 to 100 times greater than this. Even more than in atmospheric science, understanding relies on computer models, and some global-scale effects with vital implications for our future, such as the "bipolar seesaw" under which one hemisphere is thought to warm if the other cools, are still poorly understood.

Wallace Broecker of the Lamont-Doherty Earth Observatory in New York State, who made the original link between the ocean conveyor shutdown and climate change, has pointed out that our circumstances now, near the end of an interglacial warm period, are hardly comparable with those at the end of the ice age when the Dryas event took place. He believes that it would take a substantial temperature increase of around 9°F to flick the ocean switch, a rise not expected to be seen even under the more pessimistic climate change scenarios until around 2100. And Craig Wallace, then at the Climatic Research Unit at the University of East Anglia, announced that, although his simulations suggested a weakening of the Gulf Stream by 10–50 percent by this date, "global warming will continue, which means the planet will still get hotter, only slightly less so."

Recent findings suggest that the Gulf Stream may be slowing more rapidly than expected, yet the North Atlantic in general is actually warming—leaving our future climate as hard to predict as ever. Nevertheless, the 2007 report of the Intergovernmental Panel on Climate Change doesn't make a case for cooling.

What does the more distant future hold? What happened to that fateful "orbital configuration appropriate for a full Ice Age?" Well, nothing actually. It's still coming. But exactly when is anybody's guess. The uncertainties in the Milankovich cycles leave plenty of scope for argument. Some believe that the Earth has already begun to cool; it's just that we haven't noticed because of all the warming! This view has prompted climate change skeptics to esteem our emissions of carbon dioxide as this "wonderful and unexpected gift of the industrial revolution."[5] According to Bill McGuire, professor of geophysical hazards, "We should expect our planet to be plunged into bitter cold within the next few thousand years"[6]—but this was in a popular *Guide to the End of the World* published in 2002. In the same year, climatologists came to the more cheerful view that we are actually in the midst of a longer-than-usual interglacial period, and still have at least 50,000 years to go before the big freeze.

8. Our Declining Resources

A bare cupboard is a perpetual worry. In 1972, *The Limits to Growth* plotted downward graphs showing that oil and gas, gold, zinc, copper, lead, and mercury could all run out within 20 years. They didn't. Now wheat, grazing land, fish, and water are said to be in short supply. Will there be enough to go round? At the same time, there are less direct costs to human growth as we lose diversity, both cultural and biological, that holds for us a deeper, more spiritual value.

Wild Talk

"Scientists bedeviled in fight for species"
—*Boston Globe*

The newspapers have three basic stories about the diversity of nature. First, the good news: curious creature (or plant) discovered. Next, the bad news: pretty creature (or plant) on the verge of extinction. The third category comes across as more bad news: invading foreign species threatens our delicate flowers (or animals).

The good, the bad, and the ugly, but often none is what it seems. A spectacular example of the good news came along in early 2009. "Ten amphibian species discovered in Colombia," *National Geographic* announced alongside photos of some of the frogs. The frogs were indeed an addition to the roster of Central American animals, but *National Geographic* had to hedge its bets on the status of the discovery by repeatedly reminding readers that these weren't necessarily new discoveries. Rediscoveries of species previously thought to be extinct happen all the time: In 2006, a nine-spotted ladybug was spotted in Virginia, in 2008 a Sumatran muntjac deer was found in a trap in Indonesia (incidentally appearing quite unperturbed), and again in 2008 the armored mistfrog was found in northern Australia.

The bad news may be misleading too, not because it is not bad, but because it is not news. Actual extinctions may be occasion for mourning but they are hard to date precisely and are rare for well-known species. There are no obituaries for plants and animals. Instead, we are reproached with picture spreads of exotic species, their mug-shots arrayed like war-dead, sometimes with dramatic red stamps across them declaring that they are "endangered" or "threatened." "Polar bears listed as threatened species," the *Los Angeles Times* reported in 2008. More rarely, this story is generalized. "Earth facing 'catastrophic' loss of species" was a 2006 headline in *China Daily*. "Specialists[. . .]are calling for the urgent creation of a global body of scientists to offer advice and urge governments to halt what they call a potentially 'catastrophic loss of species.'"

Outrunning both of these in frequency, though, in the British media at least, are stories of invasion and conquest, where one species comes into conflict with others. Favored species tend to get written up on their way down ("Endangered birds being wiped out by grey squirrels"), those we dislike on their way up ("Moth that can kill humans is found breeding in Britain"). These articles express fears of human immigration by proxy, as a *Daily Telegraph* headline-writer could not help but reveal: "Forget the plumbers, now Poland is sending us its rare butterflies." But biology does not discriminate in this way. Whether we like a certain species or regard it as a pest, it is a species nevertheless. Invasion may seem to be good for local biodiversity, but often it is bad as the invading species can harm the survival chances of many indigenous species.

These vignettes suggest accurately enough that the balance of nature is always changing and that there will always be "winners" and "losers" among individual species of plants and animals. What they do not reliably show is the systematically increasing endangerment—and loss—of species from the Earth. In 1996 the International Union for the Conservation of Nature updated its "Red Lists" of threatened species around the world. They counted

20 percent of all mammals, for example, as "vulnerable," "endangered," or "critically endangered." When the lists were updated in 2006, 70 species of mammal had gone extinct, although those considered vulnerable or worse had held steady at 20 percent. All other classes of species from birds to insects to plants were worse off to varying degrees, with amphibians especially badly hit, rising from 2 percent threatened to 31 percent.[1]

In press stories, the threatened species is often exotic, its existence threatened by practices such as tropical logging over which we tell ourselves we have no control. The creature's predicament is to be pitied, but there's nothing we can do. It was thus an ironic surprise in 2009 when construction of the world's largest solar power plant—slated to be built in southern California—was halted following the find of a small population of red-cheeked squirrels. According to *ecoEnquirer*, "The situation has confounded local environmentalists, who are now evenly divided on whether the solar power plant or the nest of squirrels is more important to their cause."

This brings home the big questions about biodiversity loss. How great is it? Which way is it heading? Is it our fault? What must we give up so that the nine-spotted ladybug, the Sumatran muntjac deer, and the armored mistfrog may prosper? And, shockingly perhaps, does biodiversity matter anyway?

There have always been species losses, most notably during 5 mass extinction events. Best known is the probable asteroid strike on the Earth that led to the extinction of 85 percent of species including the dinosaurs 65 million years ago. The other extinctions all occurred more than 200 million years ago, including the super-volcano or asteroid that wiped out 95 percent of species 251 million years ago. However, these catastrophic but infrequent events account for only 4 percent of species extinctions over time. Extinctions occur continually due to natural selection. A species lifetime is typically a few million years, and most species that have ever lived are now extinct.

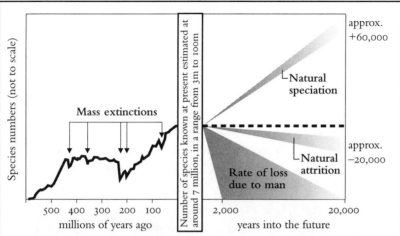

The left-hand side of the chart shows the natural trend for biodiversity to increase, punctuated by the 5 major extinction events to the present. The data point in the center shows the range in number of species thought to exist today. The right-hand side shows natural rates of both speciation and species loss as well as species loss due to man extrapolated from current estimates. The shaded areas of the extrapolated lines schematically represent uncertainty in estimates of rates of species gain and loss.

Source: Millennium Ecosystem Assessment/ Measuring Conservation for Biodiversity (Royal Society, 2003)/ E.O. Wilson, The Diversity of Life.

All these losses are more than compensated in the normal scheme of things by the rate at which new species evolve. Natural speciation is thought to give rise to 3 new species a year, although it is hard to be sure as scientists have not observed the process directly. This process has ensured that the present era of life on Earth is more diverse than at any time in its 600-million-year history.

However, the current rate of species loss is very great, perhaps 100 to 1,000 times this rate of increase, as we humans compete against other species for food and water and destroy their habitat. The "globalization of nature," by which invasive species spread with either deliberate or inadvertent human assistance, is accelerating this loss.[2] Scientists estimate that the Earth loses on average 1 species a day, including roughly 1 bird and 1 mammal every year—2007's mammal was the Yangtze river dolphin, the probable demise of which occasioned an obituary in *National Geographic*. While recorded extinctions seem tolerably low, the loss of *populations* across what an economist might term a healthy "basket of species" is much higher. WWF's Living Planet Index is such a basket, representing 1,300 land, freshwater, and marine species, and it has declined by 30 percent from 1970 to 2003.[3]

There are several good reasons to be concerned about this. Americans such as the biologist Edward O. Wilson have found that the argument that works best with Congress and organizations like the World Bank is a utilitarian one. We depend on plants and animals not only for food, but also for oxygen generation, water capture, protection against natural disasters, clothing, labor, transport, and medicines. And there is huge untapped potential among species yet to be exploited. According to the UK Natural Environment Research Council, Britons obtain 90 percent of their calories from just 30 crops, for example, but there are 1,650 species that could be grown for food—and the situation is presumably similar for Americans. The Amazon river turtle deserves to be saved, Wilson has said, not only for its own sake, but because it tastes good—and could yield 400 times the amount of meat produced by cattle raised in the same area of cleared forest. Meanwhile, the pharmaceutical multinationals are hacking their way through the jungles of Central America in search of natural drugs.

The utilitarian argument may be extended to point out that benefits to humankind may be obtained not only from individual species but from biodiversity itself. The health of the planet may depend not so much on individual species but on the interaction between them, perhaps especially involving the vast numbers of microbial species in the soil. But quite how this may be so is not understood, and scientists' opinions have differed on whether a high level of biodiversity assists the working of the global ecosystem.

Quite different is the aesthetic argument often favored in North America that biodiversity should be preserved for our pleasure. We should look after species in the same way that we look after works of art—a loss may be afforded economically, but it is always to be regretted. If we grow to love our ecosystems, then that very familiarity will help to save them, even if it does mean that we try hard to preserve the glamorous species and ignore the dull ones.

Finally, there is a purely moral supposition that humankind has a duty of global stewardship. Upon this argument may depend the survival of those unfortunate species that we find neither useful nor pretty.

The problem with these arguments is that we don't really know what we are talking about. There are probably something less than 10 million species, although estimates range hugely between around 2 and 100 million, according to the World Resources Institute.[4] Only 1.7–1.8 million of these species have names, and even this figure is uncertain because there is no central catalog and many species are certain to "exist" only as synonyms. Most plants and vertebrates are named, but there are perhaps seven more insects for each of the million with a name already, while only 100,000 of the world's estimated 1.5 million fungi species are named. Still less is known about marine micro-organisms and microbes in the soil, the numbers of which cannot be estimated by the methods used for larger species. Then, within species, even many of the large and obvious ones, there are high uncertainties as to population numbers and trends. So when the 2002 World Summit on Sustainable Development set a target for the reduction of biodiversity loss, it did so despite the fact that "[n]o sound scientific basis currently exists for assessing global performance against this target."[5]

Small wonder, then, that some species finds turn out to be previous losses (such as the nine-spotted ladybug). Others go straight from unknown on to the endangered lists. With so many species unknown, and the populations unknown of many of those we do know, it can only follow that estimates of rates of species loss and endangerment are little better than guesses. As the British Royal Society confesses, "The fate of organisms that have not yet been recognized by science cannot be measured." Extravagant claims for annual losses of species in the tens of thousands attributed to Al Gore and Ed Wilson can most charitably be explained as fractions derived from the absolute maximum possible number of species on the globe.[6]

This ignorance makes it hard to direct conservation efforts appropriately. But there is another factor that may make them irrelevant altogether. Scientists have little idea of how many species we can afford to lose, not only because of the uncertainties in their estimates of species numbers and populations, but because of the fundamental nature of evolution.

The trend in biodiversity has been an upward one for millions of years as species achieve closer adaptation to their habitats. This raises a logical conundrum with practical consequences for our efforts to contain the damage we are doing to the Earth's biodiversity. If biodiversity always tends to the maximum, it becomes meaningless to talk about an optimum level of biodiversity. And if there is no optimum level, there's less point in fretting about loss of biodiversity when this is something that only temporarily sets back nature's "aim." We cannot logically say that we can afford no loss of biodiversity, except to the extent that the loss of one species might have a detrimental impact on connected species. Of course, it would be a shame if, amid the general carnage, we were to lose one species that held the balance of a whole ecosystem, or that might have offered us some miracle medicine.

Our modern successes in agriculture, industry, and spreading our own species have so far killed off species equivalent to 100,000 years' worth of evolution. But does it matter? For utilitarian purposes, the answer is no. We need only a tiny fraction of species to support our lifestyle. Even on a higher plane, it is not clear that maximum biodiversity is necessarily a good thing—the strength of interaction between pairs of species may matter more, according to the population biologist Robert May. So our main reason to fear the loss of biodiversity may be sentimental. As the newspaper pictures remind us, we just like sharing our world with polar bears and red-cheeked squirrels.

Setting aside dilemmas over which species to save and why, the problem of conservation is a practical one. The major environmental causes of biodiversity loss are over-harvesting, pollution, climate change, aggressive species invasions, and human settlement. But they all come down to loss of suitable habitat. In the 1960s, Ed Wilson developed a "theory of island biogeography," which shows how the number of species sustainable within a given area depends on that area. The theory predicts how many species we can expect to find on an island (or other area of habitat that is in effect isolated) of a given size. It also predicts how diversity will be lost if the habitat is reduced. Roughly, if an isolated habitat is cut to a tenth of its former size, it will be able to support only half the species it used to. The theory also shows how fragmented areas that look like a sustainable habitat on a map may not suffice if the species in question will not cross from one area to the next.

This is a major reason why, even in the admission of scientists, many conservation efforts have been unsuccessful to date. The areas devoted to wildlife have been too small and too broken up. Even the vast national parks of North America are proving too fragmented for some species, whose decline, though slowed, has followed the pattern predicted by Wilson's theory.

The answer is much bigger, better-located, and better-regulated reserves. Ideally, we should focus on conserving habitat—then the species that live there will be saved automatically. But being the sentimental souls we are, we prefer to cherish glamorous species of rare orchid or the iconic panda. Fortunately, this is almost as good. If the Chinese succeed in saving the panda—despite the country's galloping industrialization, conservation efforts are doing well, and recent fieldwork has shown there are more pandas than were thought—it will be because they saved enough of its habitat, and with it hundreds of other species without really trying.

It is the same in southern California. Should we care more about the red-cheeked squirrel than about the construction of the world's largest power plant? Perhaps not. But should we then support an overly consumptive lifestyle at the risk of biodiversity? Again, perhaps not.

The Cod Delusion

"Study sees 'global collapse' of fish species"
—*The New York Times*

On November 3, 2006, it was widely reported that the world's entire commercial fishing stocks might be exhausted within our lifetime. "Study: Seafood could disappear by 2048," was how the *Chicago Tribune* led with the item. The paper neglected to say in which month of that year we'd have reached the bottom of the barrel. But it mourned:"Clambakes, crabcakes, swordfish steaks, and even humble fish sticks could be little more than a fond memory in a few decades."

Most of the major American newspapers ran variations of the story on the same day, prompted by a paper in the journal *Science*. Often they used the word seafood rather than fish stocks, bringing home the human implications as forcefully as they could. Most cited the doomsday year 2048, and some printed a dramatic downward graph of fish supplies crashing to zero.

The idea of a resource running out plays well in the newspapers, especially when there's nothing for people to see by which they might make a judgement on the matter for themselves. (It makes bad television for the same reason.) We can't see the fish in the oceans. But it was all too easy to picture the suddenly emptied supermarket counters replenished with turnips rather than turbot. Consumers can be forgiven for feeling a little confused. Hardly had we got used to the message that we should eat more fish because its protein and omega oils were good for us than we were being told to stop in the name of environmental sustainability.

The science behind the story is more subtle and leaves room for hope. In 2003, Boris Worm and his colleagues at Dalhousie University in Halifax, Nova Scotia, had reported that 90 percent of large predator fish, such as sharks, had vanished from the oceans since 1950. This was about conservation, not the human food supply; indeed if predators were down, their prey—and our food—might prosper. But Worm's new paper in *Science* shows that this is certainly not the case. Worm and his team assessed a wide array of marine ecosystems and found them to be "rapidly losing populations, species or entire functional groups"—that is to say, whole families of species.[1] Both absolute fish numbers and diversity in terms of numbers of species showed heavy declines. The authors' significant conclusion is that these two are bound up together such that over-fishing leads to loss not only of the species fished but of wider biodiversity, and conversely that reduced diversity in itself adds to the erosion of commercial fish stocks.

The results come from a statistical analysis of many previous studies, taking into account data gathered in different ways from different areas. In other words, it is not an analysis of directly collected scientific data, but an analysis of other people's analyses—what the scientists themselves call a "meta-analysis." This is not invalid at all, but it does leave the door open for sceptics to take issue, perhaps because a particular set of data is thought less reliable than others or because different data sets are not strictly cross-comparable. The papers had no difficulty finding scientists who did not support the study, one quoting a local professor who called it "just stupid."

The Dalhousie authors themselves concede that some of the data they have used may be unreliable. Figures from the United Nations Food and Agriculture Organization (FAO), for example, are supplied by member nations, some of which may distort the information they provide in what they see to be their own interests.

The assumption factored into the projection of the data into the future—that fishing and marine conservation practices will continue unchanged—is more problematic, as is the fact that the *apparent* correlation between biodiversity and fish stocks is not *proof* of a direct connection. These are standard difficulties with scientific statistics that use present-day data to try to tell us what will happen in the future. Against this, the authors can, and do, claim that if enough surveys are taken into the reckoning, and all are found to fit the overall pattern, then it would be foolish to dismiss the major conclusion.

The more glaring loopholes—chinks of light for optimists in the fishing industry—lie beyond the terms of reference the scientists set for themselves. For a start, Worm and his colleagues focus their attention on areas where there is the most detailed data, which is to say where we fish already, and on the species that are "currently fished." This seems at first like looking at a dug-out seam of coal and saying all the coal has run out. Fish are free to move within their ecological range, however, and so a loss in a monitored area can be taken to indicate a likely loss in other areas, although it is not a guaranteed loss.

Finally, the paper does not show the newspapers' simple graph of fish stocks hitting rock bottom in 2048 (the future portion of which is an extrapolation based on a business-as-usual fisheries model). Instead it shows a different set of downward curves only to 2004, representing groups of species "currently fished," which portend potentially less dire consequences.

The Dalhousie research is both ambitious in scope and gloomy in tone, but predictions that we will run out of fish are nothing new. At the beginning of the last century, Britain reacted to depletion in the North Sea by sending its fishing fleets further north. Fifty years on, this led to a series of "cod wars" with Iceland. Globally, though, the view was that there were plenty more fish in the sea. In the 1940s, even Rachel Carson, later to write *Silent Spring*, was emphasizing not man's depredations but the ocean's bounty in two books entitled *Food from the Sea.*

The global ocean catch grew from 19 to 85 million tons from 1950 to 1990, according to FAO figures, a growth factor rather greater than for the world population, accounted for by the fact that people are eating more fish than they used to. Since 1990, the catch has fallen slightly, and the "dream anticipated by policy makers in the 1970s" of a harvest of 100 million tons seems beyond achievement.[2] In 1997, the FAO warned that 60 percent of commercial fish species were fully exploited, over-fished, or depleted. By 2002, the figure was 75 percent. National commissions warned of the worsening situation in American waters in 2003 and 2004. It now appears that the world reached a point of "peak fish" around 1994, although the moment passed unnoticed, in marked contrast to "peak oil," which we may be experiencing at the moment.

Assessment of the risk of losing a favorite food is not helped by a history of antagonism between conservation biologists and the fishing industry as well as the sheer complexity of the ecology. A U.S. National Fisheries Institute spokesperson responded tartly to the Dalhousie findings: "Fish stocks naturally fluctuate in population."[3] Fishermen observe that warming oceans may be important, but this may simply be part of a denial mechanism by the fishing community to blame external factors rather than its own malpractice. There are frequent reports of exotic warm-water fish newly found in British waters, for example, while fleets have to go further north to find familiar species. But are the latter fish found in the north because they have migrated there due to climate change or because Iceland and Norway have better managed the resources off their coasts? It is impossible to say for sure.

For their part, policy makers often tread a middle line, which may be diplomatic but is perhaps not sustainable. In November 2006, for example, the International Commission for the Conservation of Atlantic Tunas cut quotas for Mediterranean bluefin tuna, but not by as much as the body's scientists were recommending, prompting environmentalists to repeat warnings that the fish would soon vanish entirely from these waters. The next year, the U.S. formally accused the EU of failing to take the quotas seriously.

How bad is it really? It seems likely that individual consumption of fish cannot keep growing as it has been doing, but there is no reason to believe that we will be fresh out of all fish by 2048. There are grounds for optimism on many fronts. Much of the increased global demand has been met by fish farming, or aquaculture, which now yields 40 million tons a year, up from just 13 million tons as recently as 1990. There are severe problems with aquaculture as it is presently practiced, merely the least of them being that the product frequently tastes like blotting paper. Using grain and fishmeal for feed is highly inefficient. It can take 3 tons of ocean fish to make meal to feed 1 ton of farmed fish. Chemical and genetic contamination of the marine environment is also an issue. But aquaculture is a young industry that is experiencing many of the historical difficulties of intensive agriculture all at once. Imagine the chaos and protest that would follow on land if acres of forest, heath and wetland were suddenly turned over to wheat and cattle. In principle, farmed fish offer a more sustainable source of protein than livestock.

It is possible, though always politically unpopular, to scale back sea fishing. Norway did this with its cod fleets in 1989 and found that stocks bounced back amazingly quickly, so that by 1992 it was possible to relax the restrictions. The short-term economic and social cost was considerable, but not as great as the cost of ignoring the problem, which is what happened in Newfoundland, leading to the extinction of cod there and the permanent loss of the vast industry they once supported. However, regulations and quota systems must be enforced if they are to be effective.

A fifth of the global catch—some 20 million tons a year—is unwanted or disallowed under quota rules and is simply thrown back in the sea dead. Reducing this by-catch is a priority. Other efficiencies may be possible. At present, the harvesting of fish is allowed to proceed in an unconscionably crude way as Charles Clover illustrates in his book *End of the Line* with a dramatic comparison of how trawling might appear on land, indiscriminately killing and mangling as well as catching all kinds of animals. Fishing is an industry, not a romantic calling, and just as technology has helped us ruthlessly to track down ever more elusive stocks, so there may be technological means of selecting fish for harvest without damaging other species or the supporting ecosystem as much as we do now.

More and larger marine reserves would work not like safari parks but like game nurseries, providing refuge where depleted species could recover to sustainable levels so that we could then continue to harvest them from unprotected seas. Such reserves would also help to maintain the biodiversity upon which population levels of economically important species depend. As Worm notes: "There is no dichotomy between biodiversity conservation and long-term economic development; they must be viewed as interdependent societal goals." Encouragingly, while fish stocks decline when biodiversity is lost, they can also recover rapidly when diversity is high, making the fisherman's job easier once more.

Relatively little is known about the global diversity of species, wherever they occur. The oceans, and especially the deep oceans, are far less explored than the land surface, so it seems probable that they harbor many unknown riches. Given this, it seems hardly likely that we have an accurate handle on the ocean's total potential for food.

There is much that consumers can do too. It is their conservative eating habits that drive up prices and bring favorite species to the brink, with cod in Stockholm at $80 a pound and sushi tuna in Tokyo at $100. Change is not impossible. Herring, not cod, was once the over-fished staple in northern Europe. The Romans paid $10 for a 1-pound red mullet, a fish for which modern buyers are accustomed to paying only a fraction of that amount. The market mechanism slows the growth in demand for such fish but does not put any block on their ultimate spiral to extinction. Here the WWF conservation charity may help by prompting retailers, even counting the environmentalists' *bête noire* of supermarkets, Walmart, to sell more sustainable fish.

Meanwhile, we consumers should consider alternative species. During the cod wars, it was suggested that the British try pollack and whiting for their beloved fish and chips, but they never caught on. If we are to keep eating fish, we will have to become more eclectic in our tastes. Some fish, such as skipjack tuna and chub mackerel, are presently conspicuously under-fished. Variety on the fishmonger's slab (and more importantly in the vast quantities of fish taken for factory processing and fish-and-chip shops) would then more closely reflect diversity in the ocean. Marine ecosystems with more species are less liable to collapse than those with few species, so that the more we spread our harvest among different species, the better the chances that stocks will be maintained of them all. The overall gain is greater than the sum of the gains made by each species because of the beneficial effect of overall species richness. Furthermore, because of the observed potential for fish stocks to recover when they are given a chance, the switch away from present favorites need only be a temporary measure.

This potential has been dramatically demonstrated not once but twice in the last century as a side-effect of two world wars. By 1945, according to Mark Kurlansky's *Cod*, "fish stocks in the European North Atlantic, after six years with little fishing, were at a level that has never been seen since."[4] A big war would do the trick again. But binding international agreement on fisheries management seems preferable.

Not a Word

"Like ancient forests displaced by houses, language is ended too"
—*The Times* (of London)

Any discussion about the death or loss of languages rapidly comes up against a fundamental problem of definitions, making them hard to count and analyze. What is the difference between a dialect and a language? And do sign languages count? The consensus figure is that there are about 6,000 to 7,000 languages currently in use globally, but estimates in the academic literature range from 3,000 to 10,000. The practicalities also mean that it is impossible to keep an accurate tally on the number of moribund, endangered, or safe languages.

Even so, the facts seem to be compelling and show a clear trend. We are told that the world is losing several languages every month and that 90 percent of the world's languages will effectively vanish from day-to-day use by the middle of this century.[1] Most languages are losing speakers, despite the rapid global population growth. At least 500 of them are considered to be nearly extinct in that they are spoken by fewer than 100 people. It has been quipped that some Native American languages are only kept alive by a few old parrots on the Orinoco River!

It is a clear sign when only the elderly and a very small fraction of an ethnic group are speaking a language that it is effectively condemned to extinction. The *Houston Chronicle* told the story of a German dialect spoken in Texas that is expected to vanish within 30 years.[2] The paper said of one elderly couple that "the language will likely die with them," as their children had not been persuaded to learn it. The adverse dynamics can also affect languages with a larger number of speakers. It is estimated, for example, that there are 500,000 speakers of Breton (a Celtic language spoken in Brittany and Loire-Atlantique, both in France) over the age of 50, but fewer than 2,000 under the age of 25, so unless action is taken to save it, the language will effectively die out in the next half-century. Even some languages used by very large numbers—such as some

Indonesian languages with a million or more speakers—are at risk due to their speakers' age profile.

The loss of languages is a marked change from 10,000 years ago when—although no one knows for sure—there could have been 1 language for every 500 or so people on the planet. The expert community estimates that there could have been anything between 30,000 and half a million languages that have come and gone without trace. Isolation and the lack of trade and transport naturally bred linguistic diversity. Now, in contrast, it is estimated that there is 1 language for every 1 million people. But, as ever, averages can lie and all languages are, of course, not in equal usage. Indeed, 95 percent of them are spoken by just 5 percent of people worldwide and perhaps as many as 350 languages, about 5 percent of the total, have more than 1 million speakers.

It is the countries with the greatest number of languages that are losing languages the fastest. Brazil is home to around 30 nearly extinct languages, and the U.S. has around 70, but it is Australia in the lead with as many as 200 nearly extinct languages. Nearly 90 percent of Australia's languages—those spoken by aboriginal peoples—are expected to perish with the current generation. A similar fate awaits many of Africa's tribal languages. Roughly one-third are said to be endangered, and it will be very difficult to save them since the vast majority have no written record.

The speed of loss is accelerating as languages become a victim of rapid globalization. New languages have traditionally appeared through pidgins and creoles, merging with other languages into families, sometimes taking on a written form. They grow and increase their influence, as did Greek and Latin, and then mutate. Accidents of history, such as colonization and trade, gave some, mainly European languages and in particular English, an importance well above their original geographical and cultural weight. But while some languages are vibrant and evolving, many more are being abandoned and few are being created, as the old dynamic has ground to a halt. When communities find that their ability to survive and advance economically is improved by the use of another language, native tongues naturally fade away, often rapidly as the young seek new opportunities. If languages are used just for religious ceremonies and bedtime stories, rather than trade and government, they are much more likely to die.

But how important is it to save languages? Some people feel this loss really matters—describing it as a disaster for humanity—believing it to be a more fundamental concern than a reduction in plant and animal diversity or the destruction of culture per se. One newspaper referred to the "extraordinary interaction between language biodiversity," alluding to the fact that native tongues can also be rich in knowledge of the environment—flora and fauna—and of traditional, herbal medicines, knowledge that could be irretrievably lost.[3]

Languages can be full of cultural knowledge that facilitates different ways of understanding and discussing the world. For example, the Australian aboriginal language of Guugu Yimithirr does not have a concept of left and right, relying instead on the concepts of north, south, east, and west. Your left hand, in other words, could be your north hand unless you are facing in the opposite direction, in which case it would be your south hand! The language requires a constant awareness of where one stands within the landscape, geographically speaking—an alertness that is utterly lost in modern speech. One linguistics professor put it more starkly: "If you lose your language, you lose a big chunk of your identity."[4] UNESCO, the education, science, and cultural wing of the United Nations, has responded to the concern by declaring the 21st of February as International Mother Language Day, in an attempt to promote linguistic diversity.

A few minority tongues have been saved and perhaps rejuvenated. In 2005, the European Union added Irish Gaelic to its list of official languages—the hiring of translators and its use in speeches will help to preserve Ireland's native tongue, but at a cost to Europe's taxpayers of about €4 million. In Kahnawacke, a small community near Montreal in Canada, the use of Mohawk has been encouraged in schools, church services, and even a local radio station. The

University of Manchester is identifying and transcribing the many Romany dialects—spoken by small groups in many European countries—aiming to preserve the endangered language. Hebrew, essentially extinct for day-to-day communication until the nineteenth century, is perhaps the most successful revived language as it is now spoken by over 7 million people—it is the official language of Israel and is studied in many Jewish communities around the world.

There would seem to be no such grounds for concern for the world's most popular languages—around 10 of them have at least 100 million speakers. Yet even among those who use widely spoken languages, it would be wrong to underestimate the concerns about the future.

The Académie Française, for example, the pre-eminent learned body on matters relating to the French language, is not resting even though the language is spoken by over 70 million (and at least double that number according to French sources!). Notably, the Académie has tried to prevent the anglicization of the French language, suggesting that words such as walkman and software be avoided in favor of words derived from French. One of its past chairmen launched a campaign to try to make French the official language of European law and said that the defense of the language should be "the major national cause of the 21st century."[5] One suspects that, in downgrading world peace, the environment and global poverty, he might not be speaking for the French population as a whole. The campaign reflected the declining international clout of French—in less than 20 years, the proportion of EU documents originating in French has been more than halved. The French might see the use of their language fading but they pointedly say that the English must protect their language if it is not to be overrun by a bastardized Anglo-American.

There are conservatives in America who believe that English should be added to the endangered list. Their particular worry is that English will not survive the "immigrant flood" of Spanish-speaking immigrants. This seems an implausible scenario, but the American continent has a long history of immigrant languages killing off the indigenous ones. They argue that it would be unfortunate if future generations of Americans were unable to read Shakespeare in the original, but that rather presupposes that many people are reading it now!

None of these problems affects the most widely used language in the world, Mandarin Chinese, which is spoken by nearly 1 billion people. Nor should they really affect Spanish and English, which are a distant second and third, with about 350 million speakers each, followed by Bengali, Hindi, Portuguese, and Russian, with between 150 and 200 million speakers.

English is not only the most commonly spoken second language but also the lingua franca in international business, media, scientific, and academic worlds. That is just as well as there are concerns that the English themselves are becoming increasingly monolingual—in an ever-shrinking world, the ability to speak several languages should be prized. In 2004 the British government decided that learning a foreign language need no longer be compulsory beyond the age of 14, and modern language departments of universities are closing as demand for places tails off, so Americans are no longer the only ones who are "behind" from a linguistical standpoint.

English-speakers might be relieved that it is their language that is becoming the world's language, but it comes at a cost as the new global language looks less and less like English every year as it absorbs approximations and distortions. As distressing as this might be to many English-speakers, such concerns are surely misplaced. While it is no doubt a bad thing to force people to stop using their language, it is hard to see why it is a bad thing if their language evolves—or even disappears—naturally. Our language is the one that we speak, that suits our needs, not the one that our ancestors spoke. Languages have always developed and expanded, withered and died, reflecting the ebb and flow of human politics, economics, and migration. They fade away with little fanfare. The many language projects are to be applauded but the realistic aim can never be higher than recording a little bit of social history for posterity.

9. Modern Science

Technology used to promise wonders. Now, it seems, it merely breeds distrust and fear. Arthur C. Clarke famously wrote that any sufficiently advanced technology is indistinguishable from magic. So why is the magic now black? Scientists insist that they are not moral agents, that their innovations can be exploited for good and ill. The public is dismayed at this abdication of responsibility and hankers for the days when scientists were heroes.

Frankenstein Foods

"Harvest of the damned"
—*Daily Mail*

On February 16, 1999, the front page of Britain's *Daily Mirror* showed a picture of Tony Blair, whom it termed in the accompanying banner headline "THE PRIME MONSTER." What had the prime minister done that was monstrous? A subheading explained: "Fury as Blair says 'I eat Frankenstein food and it's safe.'"

"Frankenstein food," for those unfamiliar with British tabloid argot, describes foods derived from genetically modified crops, such as bread made from GM wheat, or soy milk in which some fraction of the soy is genetically modified. The term has become the inevitable media shorthand for this major development in agriculture, but in 1999 it was a new coinage.

By invoking Frankenstein's monster, the critics of this biotechnology, led by Greenpeace and Friends of the Earth, and backed by the organic-farming Prince Charles, played on age-old fears that, if we mess with nature, nature will mess with us: GM food is not merely modified, but mutated, and if you eat it, you will be mutated, too. This is as absurd as suggesting that if you eat a hybridized tomato, you will become a human–tomato hybrid yourself. We eat conventionally hybridized produce all the time, and there have been no recorded instances of human–vegetable hybrids. Zeneca had launched a GM tomato paste without controversy. But vague unease was transformed to widespread public fear in 1998 when laboratory research by Arpad Pusztai at the Rowett Research Institute in Aberdeen appeared to show impairment of the immune system in rats fed with genetically modified potato. (The research was later discredited.)

Americans had, by this time, some years to get used to GM foods, but there had never been anything like the controversy seen in Europe. Yet the risks—to human health, to the environment—are essentially similar on both continents. There is no intrinsic biological difference between Americans and Europeans, and little aside from its scale between American and European agriculture. So the fact that panic in the European media was matched by indifference in the United States needs to be explained.

One factor is a recent history of biological food scares, especially in Britain, related to farming on an industrial scale, ranging from salmonella in eggs to bovine spongiform encephalopathy (BSE) in beef cattle. In the United States food safety is looked after by the Food and Drug Administration whereas in Britain at the time it was overseen by the ministry also responsible for agriculture, producing a clear conflict of interest. But perhaps the major distinction is cultural, to do with the closer connection felt (rightly or not) to exist in Europe between the field and the table, an orthodoxy to which both picturesque countryside and culinary tradition are intimately bound.

This bucolic scene seems to be in stark contrast to the vision of agriculture opened up by biotechnology, a young and optimistic industry eager to show its potential. The mid-1990s saw the successive introduction in the U.S. of the FlavrSavr tomato, an insect-resistant corn, and a herbicide-resistant soybean. Monsanto's insect-resistant corn promised to save farmers insecticide, while the herbicide-resistant soy would help the manufacturer sell more of its Roundup weedkiller. In 1996, just 1.7 million hectares were planted to GM crops. The figure rose to 11 million hectares in 1997, 28 million in 1998, and 40 million by 1999, an area the size of California. This astonishing growth continues at 10–20 percent a year and has spread from North America to Argentina, China, and a number of other non-European countries.

Britain's anti-GM campaign shifted into high gear when the *Daily Mail* launched a "Genetic Food Watch" campaign in January 1999 and began using the scare label "Frankenstein food" on its stories. Consumer attitudes to GM technology hardened, and the supermarkets judged that greater commercial opportunity lay in organic produce, which it could sell at a huge premium, than in food required to bear labeling revealing GM contents that nobody wanted. By 2003, farm-scale government trials of herbicide used on GM rape and sugar beet indicated that widespread planting would be likely to damage the environment, and the campaign seemed to be over. "The death knell sounds for GM," crowed the *Daily Mail* as Monsanto shut down its operations in Britain.

In the United States, meanwhile, GM food ingredients have become widespread without labeling regulations. Any European who has visited the country during the last decade will almost certainly have eaten GM food—an irony apparently unnoticed by the *Daily Mail* as it promoted Florida holidays to its fearful readers. Very occasionally, the American press has sided with the environmentalists, but the more general reaction has been bafflement and hurt that other countries don't want their hi-tech chow. "Food that starving people won't eat: poor countries foolish to turn down genetically modified produce," was how the *Chicago Tribune* responded when the Zambian government rejected U.S. food aid offered in the form of GM corn.

There are many reasons why people might be ambivalent toward GM food, some of them more rational than others. This transformation of agriculture raises environmental, economic, social, and ethical issues, but the most pressing concern is always human health.

Yet health may be the least of it. In many instances, no GM substance would even enter the body. Sugar might be made from GM beet, for example, but the process of refining the crop into the pure and simple chemical substance we know as sugar eliminates all biological material. In any case, our stomachs are no strangers to modified foods. Agricultural produce has been "modified" by selective breeding for thousands of years and by the deliberate creation of hybrids for nearly 200 years. A poll quoted in the ever-vigilant *Daily Mail* suggested that nearly three-quarters of people in Europe remain "extremely wary" of GM food and prefer "natural crops." In fact, if they were given "natural crops" to eat, they would quickly become both unhappy and possibly unwell.

According to the *Mail*, people's main fear is "the lack of data proving that the 'Frankenstein foods' are safe to eat"—something that can't, of course, be proven in the way that the paper seems to demand. One death from eating GM food would suggest (but not prove) that it is unsafe, but even the continued survival of 300 million human guinea pigs routinely eating the stuff in North America does not prove conclusively that GM food is safe. Nevertheless, it is the clearest evidence available that GM foods are safe in terms of human health.

Where the GM revolution is opposed in America, it is opposed less on health grounds than on environmental grounds. In 1999, a laboratory experiment at Cornell University showed that caterpillars of the monarch butterfly were damaged by consuming the pollen of corn genetically modified for insect resistance. The monarch butterfly is brightly colored and migrates in huge flocks, making it something of an American icon. The research therefore excited unusual public alarm much as Rachel Carson's singling out of the symbolic bald eagle as a species at risk of "collateral damage" from DDT had done in *Silent Spring* 36 years earlier. However, subsequent field experiments showed that the caterpillars mostly did not eat the GM pollen anyway. The danger to such species is in any case surely dwarfed by that from the conventional pesticides that GM crops would help to banish.

The monarch butterfly notwithstanding, the American farming environment is different from that of Europe, occupying vast tracts of land with inevitably depleted biodiversity. Europeans have developed a fondness for their cozier mixed countryside, which may lead them to forget that agriculture always has a detrimental effect on the natural ecology—that, in a sense, is its purpose. There remain unresolved issues to do with the invasiveness and persistence of GM varieties and their effect on other species, but GM crops may turn out to be no worse than conventional ones and, if planted appropriately, could be of comparative benefit to the environment.

The ethics of GM technology are harder to resolve—like the ethics of any commercial transaction between unequal actors. America complains that Europe is restricting free trade not only by refusing to import its GM foods directly, but also by banning produce that may have become contaminated with altered stocks, including those from developing countries. As a result, those countries feel driven to refuse to take American GM seed even to feed their own people, as happened in Zambia. Critics say this argument is self-serving, that food is anyway in surplus, and that hunger can be better alleviated by addressing distribution. Light is seldom shed when a technical matter is raised to a "moral" issue, a process eagerly assisted by both the biotechnology industry and the anti-GM lobby with their opposing messages of "feed the world" and "keep nature pure."

The company at the forefront of the GM revolution, Monsanto, furthermore acted with what in hindsight seems extraordinary carelessness, completely misjudging the European market and failing to advocate a precautionary approach to matters such as the mixing of GM and non-GM soybeans, which would later make it impossible to introduce food labeling and consequently difficult for American farmers to export their crops. While the anti-GM lobby undoubtedly crusades with "overtones of moral fanaticism," the pro-GM industry is not exactly short of self-righteousness.[1] The perception of corporate arrogance in turn allows environmental campaigners to present themselves as "people's champions tackling giant American Goliaths" despite their sometimes illegal and destructive actions.[2]

In the developed world, even the basic utilitarian argument for GM food is weak. A report to the British government in 2003 could find no discernible advantage for consumers and "limited" economic benefit.[3] This lack of benefit—admitted on occasion even by pro-GM interests—suggests that GM food may simply be a solution in search of a problem. It is an innovation that has come about because the technology exists, not because the market demands it.

This impression was heightened by scientists accustomed to their role as heroes of the so-called "green revolution" that had seen agricultural productivity raised enormously during the twentieth century through improvements in chemical fertilizers, plant and animal breeding, and sowing and harvesting technology. In genetic modification, they thought they were simply continuing in this tradition, using the latest science to make these processes more exact and efficient. The public, on the other hand, perceived a fundamental step change. Scientists thought they were aiding the supply of food worldwide; consumers suspected a corporate profit motive. Scientists saw no other way forward; consumers saw organic produce as a viable alternative.[4]

Scientists accordingly took the benefits of GM technology as self-evident and focused almost entirely on the objection to it which they were best prepared to deal with—namely the possible risk to health. Statements such as that by Peter Lachmann, who produced the Royal Society's 1998 report on GM food, that "the public furor about health hazards of genetically modified foods rests on no reliable evidence base and falls little short of mass hysteria" thus miss the point (possibly deliberately) for a public whose view of the issue was always more complicated, involving attitudes toward America, corporate monopolies, the countryside, and so on.[5]

For the crux of British consumers' fears so adroitly exploited by the *Daily Mail* was never health or the environment or ethics.

It was instead to do with perceived coercion. In 2004, it became apparent that the newspaper's campaigning had not succeeded after all, as a European Union moratorium expired and plans for new GM trials were announced. "So we're going to be force-fed GM," the paper announced. A month later, it was: "Frankenstein food? You'll be made to like it," a strange headline for a report that was actually about GM animal feed—perhaps the paper's readers were more bovine than anybody had thought.

Today, we are faced with a situation where GM technology exists, works for some, and is unlikely to go away. In February 2006, the World Trade Organization found for the United States, Canada, and Argentina and against the European Union on the conduct of review procedures for GM crops. It ruled that there could be no more unreasonable European moratoriums. The *Wall Street Journal* welcomed the outbreak of "Frankensense."

Undeterred, the *Daily Mail* now warned that "U.S. biotech firms" were going to "blitz" Ireland with "GM grapes, apples, bananas, vegetables, and cereals." Surprisingly, the article did not describe the Americans' secret plan to drop this provender from over-flying B-52s. A group of Irish chefs had banded together to resist the invasion. Now, you wouldn't necessarily ask a scientist to cook you a gourmet dinner, nor would you generally ask a chef to spell out the dangers of scientific innovation. But again this did not stop the *Mail*, which quoted one of the restauranteurs making the nonsensical claim that soon "it will be impossible to cultivate indigenous crops as GM seed is used in farms across the country."

"The U.S. government last night claimed victory in a battle to force genetically modified food on to the dinner tables of Britain and the world," was how the *Mail*'s story began. In Ireland, as elsewhere in the EU, foods with GM ingredients must be labeled, but the *Mail* immediately raised a new fear that "the U.S. administration is considering bringing a second legal case to the WTO to get this abolished."

Where will it end? The companies have learned from Monsanto's mistakes and now threaten to use GM organisms cunningly disguised in products that people might actually want. One recent innovation is a low-fat ice cream made using a protein derived from a species of fish. As a perverse way of dealing with the problem of obesity, this concoction might seem to exemplify GM's predicament as a solution looking for a problem, but the market appeal of such an idea cannot be ignored. Anti-GM campaigners sense that, as the *Times* put it, "manufacturers may be trying to introduce GM processes by stealth in the hope of making GM foods acceptable to consumers." Once they're acceptable to consumers, of course, the game's up. No wonder the environmentalists are worried.

It seems certain that the future on both sides of the Atlantic, and elsewhere, must be one where GM, conventional, and organic foods all have a share of the market. Then nobody can complain, neither the North American companies, which currently claim to be "losing" sales worth $500 million essentially because they have been unable to persuade Europeans to accept their wares, nor the crop-slashing Greenpeace and its allies, who will be seen as irrational zealots if the rest of the consumer population can calmly exercise its option to reject GM in the shops.

Little Wonder

"Fears grow that tiny particles may pose major health risks"
—*Seattle Times*

Who would be afraid of a technology that promises so much? Cars that don't need washing; lighter, stronger tennis racquets and golf clubs; bouncier balls; fridges and washing machines that eliminate bacteria; stay-sharp razors; perpetual air fresheners; stain-resistant clothes; and self-cleaning socks. And that's just in consumer goods. In cosmetics there will be more effective anti-aging creams and invisible sun lotions, in medicine, tiny diagnostic sensors that circulate in the bloodstream on the lookout for trouble, and drugs against cancer that go straight to the source of the problem. Improved foods will include milk that tells you when it's about to spoil, cholesterol-blocking cooking oil, and animal-free meat.

These wonders—and many, many others according to its enthusiasts—are the bounty of nanotechnology, a catchall term for a set of emerging techniques that may allow us to arrange atoms and molecules on the tiniest scale to make substances and objects that can do jobs that it is presently difficult or impossible to do by conventional means.

Nanotechnology first reached the public consciousness with the 1986 book *Engines of Creation* by Eric Drexler, although it was the physicist Richard Feynman who first considered the possibilities in a 1959 lecture titled "There's plenty of room at the bottom."[1] Drexler rhapsodized in more detail on the boundless possibilities of the "next revolution" in technology, as well as colorfully describing some of its pitfalls—coining the term "grey goo" as he did so.

Nanotechnology is simply the technology of doing things at the bottom of the scale—typically less than 100 nanometers (a nanometer is 1 billionth of a meter), or the distance spanned by about 1 thousand atoms. Many materials start to behave in different ways at this scale because their surface area is greatly increased. For example, a millimeter crystal of salt could be broken into 1 trillion nanoparticles with sides of 100 nanometers. These particles would have a total surface 10,000 times that of the original crystal and, if placed on the tongue, would release flavor faster.

Sun lotions work because grains of white zinc oxide reflect the sunlight and prevent it from being absorbed by the skin. Ordinary lotions use relatively large grains of oxide, which is why they appear white. But lotions recently introduced on to the market use nano-particles. These still reflect ultraviolet but not the visible light, so appear transparent. Elsewhere, too, arranging familiar materials in more precise ways will allow their potentially attractive properties to be maximized. A nanometer-smooth surface would not allow dirt or water to stick, for example. It is not hard to imagine the applications that could follow from such extensions of material properties.

Nanotechnology offers several instructive points of comparison with the GM controversy. The simple Luddite view is that we are once again meddling with nature in a fundamentally new way. "If you think GM is scary," burbled one *Daily Mail* headline: "Chicken that tastes of anything you want. Drinks that change colour at the flick of a switch. Everlasting chocolate. This is the new genera-tion of Frankenfoods being created by scientists. The consequences could be terrifying." But the argument is more subtle than this. The GM debate pitched a Europe where several countries had new left-of-center governments and heightened consumer expectations for the environment against a technologically progressive America, but nanotechnology has mobilized skeptics on both sides of the Atlantic. "Fears grow that tiny particles may pose major health risks," warned the *Seattle Times* in December 2005. The American media took the broad view that development of the technology ought to proceed, but only with regulatory safeguards. "Stricter nanotechnology laws are urged; report warns of risk to public," headlined the *Washington Post* in January 2006. The article added: "Current U.S. laws and regulations cannot adequately protect the public against the risks."

In April, the Food and Drug Administration weighed in after learning that dozens of people had suffered from breathing problems in Germany after using an aerosol cleaning product called Magic Nano. The product was quickly withdrawn from sale. The following month, the Wilson Center, a Washington think tank, released a list of 231 products whose manufacturers advertised using nanomaterials or nanotechnology, sparking further alarm.[2] The British *Daily Mail* picked up the story. "'Hidden danger' in anti-ageing cream," it shrieked. Nanoparticles in products such as L'Oréal's Revitalift, Lancôme's Renergie and Boots' Soltan sun lotion could cause "untold damage to human health," it warned, listing chest complaints, heart attacks and cancer among the problems that might arise.

By the end of the year, the U.S. Environmental Protection Agency was in on the action too, with plans to regulate the silver nanoparticles used for their bactericidal properties in domestic appliances and food containers. The *Seattle Times* noted these would be the "First federal restrictions on new, growing technology."

Even the technologically gung-ho Bay Area grew fearful. The *San Francisco Chronicle* hailed Berkeley's aim to impose "the world's first local regulation of nanomaterials." This would be aimed at the city's famous university and the Lawrence Berkeley National Laboratory, and at start-up firms that might spin off to commercialize their innovations. Researchers responded by pointing out that their laboratory nanomaterials were made in quantities too tiny to test.

In another notable difference from the GM story, scientists have joined the call for greater regulation, desperate to avoid having what they see as another potentially exciting and lucrative technology disappear down the plughole of ill-informed public fears. "The spectre of possible harm—whether real or imagined—is threatening to slow the development of nanotechnology unless sound, independent and authoritative information is developed on what the risks are, and how to avoid them," fourteen scientists wrote in the journal *Nature*.[3] Countries with large hi-tech sectors such as the United States and Switzerland have sponsored preemptive programs of public dialogue. "The development of nanotechnologies has become an ideal testing ground for the application of public engagement processes to science and technology," proclaimed Britain's Nanotechnology Engagement Group, steeling itself for the battle ahead.

Less clear was what should actually be regulated—the technological processes, which are not exposed to the public, or the materials, many of which are already regulated at the macroscopic scale. The obvious answer emerges if we look at nanotechnology not as Drexler's "coming era" of miraculous novelty, but for what it is.

Now, what follows may allay your fears about nanotechnology, or it may simply confirm them: Think of nanotechnology not as a fundamental novelty but as chemistry rebranded. Nanoparticles are simply chemicals, and, as such, to be feared or exploited according to their properties. When a government scientist warns that "it's been shown that free nanoparticles inhaled can go straight to the brain," he's omitting to say that thousands of "ordinary" chemicals do this too.[4] Prince Charles—a key opponent of the technology—spoke of "technologies which work at the level of the basic building blocks of life itself," as if these were entirely new—and sinister.[5] But nanomaterials are not new, nor are they inherently sinister, or for that matter especially wondrous. Ordinary chemicals work at this level already. The banality of the products on the Wilson Center's list attests to that. In short, the hazards of nanomaterials may be treated by and large like other chemical hazards.

The more distant dream, or nightmare, of tiny robots using atoms and molecules as building blocks sounds worrying until you realize what an unbelievably difficult way of doing things it is. Nature has taken millions of years to get to the point where tiny cells are able to grow and divide. Thousands or millions of cells must do this continuously to grow a living organism. Yet to build one simple virus, to take a Crichtonesque example, atom by atom with each assembly operation taking just a second would take a week. Man has long found it more convenient to make things by manufacturing the chemical ingredients in bulk and then shaping them into parts which can be assembled into a product. The nanotechnologist's dream is to turn this process on its head: your home might have a kind of programmable manufacturing appliance into which you would simply load some generic raw material and out would pop the latest model of cell phone, or a Prada handbag, or a pristine copy of *Prey*.

The central fear about nanotechnology is that there are, as the *Observer* put it, "no new rules." But need there be? Old rules may suffice in many cases. Beneath the noise of all the breast-beating, this is in fact how the regulatory drive begins to sound. Berkeley's nanotechnology ordinance, now approved, simply requires companies to disclose the toxicology—to the extent that it is known—of nanoparticles as they must already for bulk hazardous materials.

Some nanoparticles may indeed prove to be harmful—it would be a surprise if none did. Nanofibrers may behave like asbestos fibers. Nanomaterials slathered on the skin may be absorbed and have harmful effects. Like diesel exhaust particulates, nanoparticles in the air may enter the lungs. A more immediate concern may be the escape into the environment of hard-to-recover silver nanoparticles from discarded appliances. "Real defects" do need to be addressed, and this technology, like all others, will have to be brought under "social control." It increasingly looks as if the way this might be done is when somebody brings the first lawsuit.

Perhaps then we will know where to direct our distrust. When the Magic Nano cleaning product was withdrawn from the market, American newspapers reported on German officials' caution in assessing the problem. The assumption was that the cause of the alleged breathing difficulties was the aerosol vapor produced when using the cleaner, although "they could not rule out whether the nanoparticles it contained also contributed."[6] After tests by the German Federal Institute for Risk Assessment, however, Magic Nano was found to contain no nanoparticles at all. Consumers should perhaps worry less about new technologies and more about trading standards.

Exposed

"Nuclear nightmare"
—*Time*

In so many ways, it was made for the media. A former Russian spy, Alexander Litvinenko, is admitted to a hospital exhibiting symptoms of poisoning following a visit to a London sushi bar. At first, the speculation is that he has been given thallium. Then X-rays seem to show pieces of "dense matter" inside his body, perhaps a canister that has broken open. Finally, his condition undergoes a sudden deterioration and he dies, but not before preparing a statement blaming the Russian president, Vladimir Putin, himself previously head of the nation's secret service, for his death. The autopsy confirms that Litvinenko was killed by a large dose of radioactive polonium, a substance only likely to have been obtained with state sanction from a nuclear reactor. "NUKED," screamed the front page of the *Daily Mirror* the next day, a Saturday.

At this point the story abruptly changes gear. Now it is no longer just an intrigue. It is all of a sudden a "public health scare."[1] Over the weekend, the UK Health Protection Agency urges people who had been to the sushi bar on the day in question to contact a telephone helpline, though doctors insist they aren't at any serious risk. "Hundreds of people face being tested for traces of deadly polonium," the Sunday *Observer* writes. By Monday, nearly 1,100 people have come forward. Eight are referred for tests—"at a secret London clinic."[2] The sushi bar is closed down for decontamination as other locations Litvinenko visited test positive for polonium 210. Traces of radiation are found in other places, including British Airways planes on the Moscow route. By the end of the week, the number of calls from the public is nudging 3,000; 24 are referred for tests.

What made it such a perfect story? The suspicion of foul play—the conclusion drawn by British official reports. The exotic poisons—thallium features in a novel by Agatha Christie, while polonium cost its discoverer, Marie Curie, her life. The story unfolded in a gripping way, with each new morsel eagerly snapped up by a public with an appetite already sharpened for such incredible tales by the opening the same week of the latest James Bond film, *Casino Royale*.

Above all, there was the dread word "radiation," accompanied on the BBC news and in the tabloids by its spooky mandala. And, significantly, there was also the probability that, while everyone was excited by the drama, no reader was really at risk. This is because polonium 210 emits alpha particles, which have high energy but very short range. In Litvinenko's body, the particles damaged his internal organs over a period of weeks, leading him to exhibit the series of puzzling symptoms that kept doctors guessing as to his illness, but could not penetrate beyond, explaining why radioactive poisoning was not identified sooner. He was finally buried in what the BBC *Newsnight* program described reassuringly as a "radiation-proof coffin," although an ordinary pine one would have done just fine.

Alpha radiation does not spread through the environment. Contamination is therefore very limited. In the end the only people dangerously exposed to polonium were involved in the world of espionage. Sixteen other people received radiation doses of "some concern," including some of the bar staff, where it is thought that the poison was administered. These people face a "very small" raised risk of cancer in the long term, roughly equivalent to that if they were to live in Cornwall for a decade or two and be exposed to the high natural level of radiation from radon gas released from the local geology. There never was a risk to general "public health."

The Litvinenko episode was unprecedented in many ways, but it was not the only recent radiation scare. At the beginning of 2004, it was reported that American cities were being scanned for terrorist "dirty bombs"—bombs combining radioactive material with conventional explosives that would spread radiation. In an attempt at reassurance, the newspapers conjured the unintentionally hilarious vision of scientists, "dressed casually to blend in with people enjoying the Christmas and New Year holidays," roaming major American cities with golf bags weighed down with Geiger counters.[3] Yet they would have detected nothing if the poisoned Litvinenko had walked right past them in the street. Extravagant precautions such as this, and the HPA's offer to "worried" members of the British public to call the hotline, are signs of a desperate urge to reassure far more than they are realistic means of damage limitation.

Radiation generally invokes fear regardless of its nature, origin, or potency. During the polonium scare, the *Observer* noted that "The last radioactivity incident of this magnitude occurred when the radiation plume from Chernobyl swept over Britain in 1986." Yet that event was quite different in all of these qualities. Why does radiation inspire such terror? In *The Perception of Risk*, Paul Slovic suggests that part of the fear stems from "transmutation."[4] Polonium is transmuted to lead when it emits its alpha particle, for example. Like the mutation of viruses that also makes us uneasy, transmutation is change of a kind we cannot readily comprehend. But what we fear most is surely the knowledge—one of the few things we do know about it—that radiation causes cancer.

This is true both of nuclear radiation, which involves the emission of particles or rays from radioactive material, and of some forms of electromagnetic radiation. Ignorance of the precise facts, however, means that we are indiscriminate when it comes to assessing radiation risks. We are apt to judge the risk not according to its physical nature but according to the context in which we experience it. Thus, we judge X-rays safe (which experts warn is not the case) and the hazard from nuclear power plants high when experts insist it is low. We may be agitated when manufacturers irradiate our food in order to destroy bacteria, but we regard radon leaking from the ground into our homes with "apathy" because it is a natural occurrence.[5]

Excessive reassurance may be the contemporary response to potential nuclear radiation hazards, but in the past it has often been excessive denial. This provides one further reason why our fears persist. Early nuclear accidents at Windscale (in the UK, later renamed Sellafield) and near Detroit were downplayed in a time still marked by collective optimism about a clean, modern source of energy. The meltdown of a reactor and release of radioactive gas in 1979 at Three Mile Island in Pennsylvania was less serious than either of these events, but by this time the environmental movement had sprung up, highlighting the issue of radioactive waste disposal as well as plant safety, and the tide of public opinion turned.

This accident was overshadowed on April 26, 1986 when a reactor failed at the Chernobyl nuclear power plant in the Ukraine, leading to a chemical explosion in which much of the 190 tons of nuclear fuel escaped into the atmosphere. The radioactive rain that drifted westward over Europe and beyond contained perhaps 100 times the fallout from the Hiroshima bomb. The disaster put the brakes on nuclear power globally and is still the cause of massive controversy.

Did just fifty people die as the International Atomic Energy Agency (IAEA) and World Health Organization (WHO) insist? Or was the figure 10,000 times that, more like half a million, as is now claimed by authorities in the Ukraine, a figure many times greater than maximum estimates given elsewhere?

The World Nuclear Association, an industry lobby group, puts the Chernobyl death toll at "31+," limiting the count to those who died in the immediate blast or from acute radiation within a year and omitting all subsequent radiation deaths. It thoughtfully offers a comparison with deaths in coal mines since that date, which add up to more than 3,000, mostly in China, with another 2,000 or so deaths related to oil and gas extraction.[6] But the IAEA/WHO calculate that deaths from thyroid cancer due to radioactive iodine and other causes directly attributable to the Chernobyl explosion will ultimately be comparable with this total, at somewhere over 4,000.

Chernobyl is now a destination for day-tripping tourists, but, further west, it continues to leave its taint. The British government first denied there was a problem with radioactive fallout and then put in place restrictions on the sale and movement of sheep that were grazing contaminated grass. Eventually, the restriction zone was narrowed, oddly enough to a small area "just downwind from the huge international Sellafield nuclear reprocessing complex," a site already notorious for radiation leaks.[7] According to Brian Wynne of the Centre for the Study of Environmental Change at nearby Lancaster University, scientists originally claimed that the radiation was entirely due to Chernobyl, but later admitted that half of it came from "other sources," namely Sellafield itself and atmospheric weapons testing. Today, sheep graze the contaminated grass without restriction.

Public response to electromagnetic radiation is less understandable. Electromagnetic radiation includes radio waves, microwaves, infrared, visible light, ultraviolet, and X-rays. Only the last two of these are sufficiently energetic to produce the chemical changes in biological molecules necessary to cause cancer. Yet many people are more worried about the low-energy electromagnetic radiation from electrical appliances.

This fear was fueled by a series of *New Yorker* articles on microwaves by Paul Brodeur, expanded in a book called *The Zapping of America* (1977). Brodeur returned to the magazine in the 1980s to give power lines the same treatment. On both occasions, he built his case on anecdotal evidence of "clusters" of cancers, miscarriages, birth defects, and childhood leukemia.

Misleadingly using the terminology of *nuclear* radiation, Brodeur claimed that levels of electromagnetic radiation in New York City were a hundred million times the "natural" background. This is like pointing out that light levels in Times Square at night are this much higher than "natural." The abundance of something harmless is of little consequence. Nevertheless, the widespread scare that Brodeur's articles provoked nearly stifled the infant microwave oven industry, while public protests prevented the National Weather Service from installing a weather radar on Long Island. As Robert Park put it in *Voodoo Science*, "people feared the known dangers of howling wind and crashing ocean waves less than they feared the unproven hazard of silent, invisible microwaves."[8] The adjectives are significant here: it is the familiar, palpable danger that we downplay in favor of the novel and imperceptible.

Extensive reviews of the evidence by the U.S. National Academy of Sciences in 1996 and the National Cancer Institute the following year found no correlation between the claimed cause and effect. Even a "cluster" of five children with leukemia at the heart of the controversy was too small to have significance—it was random bad luck, not a correlation with anything. The two reports should have put the matter to rest. However, the story persists.

The fundamental flaw in the argument of those who believe that fields of any sort are the cause of their complaint is something called the "inverse square law." This says that if you double your distance from a radiation source your exposure is reduced by a factor of four. So if you worry about the 10,000 volts surging through the power lines 10 metres overhead, you should also consider the 240 volts in the cable running by your desk and the 12-volt electric toothbrush in your hand.

It seems that some people are doing just that. "Electrosensitivity" is a growing complaint. Sufferers experience headaches, skin irritation, and fatigue. The symptoms may be real enough. But have they correctly identified the cause? Or is it a case of mistaken attribution as with some claimed allergies? As with power lines, much expensive research has been commissioned to find out.

The current focus of the debate is mobile phone technology and the suspicion that the microwave signals that pass between the hand-held receiver and the nearest phone mast may cause brain damage. The argument is superficially plausible because people hold the devices against the side of their head. But again, as with power lines, no evidence solidly supports a connection. It is impossible to prove a negative, however, and so doubts persist, and, as the *Financial Times* warned, with around 2 billion users, "even a tiny individual health risk could translate into thousands of deaths."

The U.S. Department of Health and Human Services concluded that "although some studies have raised concerns, the scientific research, when taken together, does not indicate a significant association between cell phone use and health effects." Good news, then, both for the complacent and the worried. The evidence comes from small-scale studies involving volunteers, and the cause of the effects is unclear. In a sign of the times, the DHHS nevertheless advised a precautionary approach, echoed by the WHO. Such advice shows how official agencies are increasingly taking into account not only scientific evidence but also the vagaries of public opinion, evidence-based or not.

A larger study by the Karolinska Institute in Sweden has since shown that long-term phone users appear to have a greater chance of developing a particular nerve tumor. What made these results compelling was that the tumors appeared on the side of the head where the user holds the phone. But again, cause and effect have not been unequivocally linked, and other large studies have not replicated the results. The effect may stem from older models of phones that are no longer in use, or from some other factor entirely.

When computers began to enter the workplace in the 1980s, fears arose concerning the radiation from their visual displays. Research then—much of it commissioned by trade unions and coming from socially liberal countries such as Canada and Sweden—led to sensible measures to improve workplace design and office lighting and to ensure that workers took regular breaks, although no radiation hazard was ever established.

Computers were often imposed on workers who feared the loss of their jobs, but people mostly love their cell phones. So this time around, we might expect radiation fears to subside sooner than we think. Nevertheless, the WHO estimates that $200 million has already been spent worldwide on research into the supposed dangers of cell phones. As their usage expands, so the fears become more far-fetched. "Do mobile phones kill sperm?" demanded one recent headline, reporting news that the Reproductive Research Center at the Cleveland Clinic in Ohio found that men who used cell phones had lower sperm counts.

This study nicely illustrates some of the problems that arise with this sort of research. The sample comprised 364 men undergoing infertility diagnosis, hardly a typical group. They were divided into groups with high, medium, and low sperm counts, and it was found that those who used a cell phone for more than four hours a day were disproportionately likely to be in the low group. Ignored was the possibility that people with such a curious lifestyle might exhibit other traits which could better explain the results—poor diet, posture, or working environment, for example. Furthermore, the researchers appear to have forgotten the inverse square law. Unless you have ears in your crotch, the radiation your testicles receive while using a mobile phone is less than from listening to your stereo. One other thing would explain the findings—talking balls.

10. They're Coming to Get You

What you don't know can't hurt you, that is what they say. The corollary must be that increased knowledge raises the capacity for endangerment. As we learn more about asteroids, there is more to fear. Sadly, the reverse does not seem to apply. We know nothing about alien life, but we fear this too. Why are tales of encounter with extraterrestrials so often couched in terms of threat? Wherefore the human predisposition to fear? Aliens might be able to show us a good time. What use is higher intelligence otherwise?

Expecting Visitors

"Apollo astronaut, Edgar Mitchell, claims UFO cover up"
—*Huffington Post*

In one of the most dramatic UFO sightings of recent years, a group of United Airlines pilots and other employees on November 7, 2006 reported seeing a gray saucer-shaped object hovering over one of the terminals at Chicago O'Hare airport before it shot up through the clouds. One of the group notified the air traffic control tower, but the controllers saw nothing, and nothing appeared on their radar. The airline and the Federal Aviation Administration said there was no further action to be taken, leaving the workers "upset that neither their bosses nor the government will take them seriously."[1]

The episode illustrates several traits common to the more plausible UFO sightings. It has apparently credible witnesses, here in the form of trained personnel accustomed to aerial observation. It was a sighting by a small group, not a lone individual or a large number of people spread over a wide area. Typically, their feelings are hurt when they are not taken seriously by officialdom. Finally, of course, there is the overwhelming likelihood that the object, although "unidentified," was not an alien spacecraft but some highly localized weather phenomenon or a piece of debris held aloft in the breeze—or a hoax.

After all, it doesn't take much to produce a phenomenon worthy of investigation. A few months before the O'Hare flying saucer, the British Ministry of Defence leapt into action to investigate the sighting by several independent witnesses one balmy summer's night of a mysterious pattern of lights hovering above Seaham, County Durham. The effect turned out to have been produced by party lanterns which had escaped into the sky like hot-air balloons. The party-givers later reported their doings: "Our garden lanterns started a UFO scare." The news item added that the lanterns are now sold with a warning that they have been mistaken for UFOs.

With sightings like this, it is hardly surprising that UFO stories tend to be regarded by the press as little more than an amusement. But the frivolity disguises a sense of unease at the possibility, however remote, that one day a flying saucer may indeed turn out to be a spacecraft containing an alien intelligence. This possibility excites curiosity in many, anticipation in some, foreboding in others, and in a few, real fear.

It was ostensibly in an effort to assuage such public fears that the United States Air Force set up a program to log UFO sightings known as Project Blue Book. Between 1947 and 1969, the military authorities examined 12,618 UFO reports. Recent British Ministry of Defence figures record 714 sightings in the six years from 2001 to 2006, about the same rate of sightings per head of population as in America.

There is nothing disreputable about seeing a UFO, or unidentified aerial phenomenon, as the official jargon now prefers to put it. They are after all "unidentified." What is peculiar is to insist that a UFO is an alien spaceship when many other explanations are far more likely.

Many daytime sightings can be put down to unfamiliar aircraft and birds, weather balloons, satellites, odd cloud formations, or other effects. Nighttime sightings of luminous shapes and patterns may be aircraft lights, reflections of ground lights from clouds, fireflies, meteors, or luminous discharges in the atmosphere (a range of phenomena largely unfamiliar to the public among which even the "well-known" auroras borealis and australis have been seen by relatively few people). Perhaps 80 percent of all UFO sightings may be quickly placed in one of these categories. When a sighting cannot be explained so readily, the question is whether the alien spaceship explanation is *definitely more likely than any of these or anything else*. In other words, which is more likely: that a bunch of airline employees see, let's say, a piece of gray polythene borne aloft from a Chicago building site, or that they truly see a flying saucer?

Where a physical phenomenon cannot be identified, a UFO sighting may be put down to a hoax, hallucination, or delusion, depending on the number of people who saw it and their circumstances. Again, the balance of likelihoods argument applies. Is it more likely that a bunch of airline employees choose to perpetrate a hoax, or that they see a flying saucer?

Of the 12,618 Blue Book UFO sightings, only 701 remained officially unexplained, meaning that they could not be firmly attributed to a physical or psychological cause. Most turned out to be due to America's own spy planes—hence the military interest in the field at all. Of the 714 sightings that came to the attention of the British Ministry of Defence in 2001–2006, 12 were "deemed to be worthy of further consideration," according to a defense minister answering questions in the House of Commons. Questioned under the Freedom of Information Act, the ministry wouldn't say what made them special, or what further consideration they actually received.

This is not to say that there are no other rational explanations for such anomalies. In one year, 1952, there was a freak level of 1,501 sightings logged by Project Blue Book, 3 times the average; 303 of them were "unexplained," 10 times the usual number. The fact that

Life magazine ran a major feature on UFOs in that year, and that a film came out of *War of the Worlds*, in which aliens come to earth aboard a glowing meteor, may not strictly count as explanations, but they certainly cannot be counted out as factors in the result. And the fact that a disproportionate number of the sightings for that year remained unexplained may in turn be simply down to the fact that they were not properly investigated, as the Air Force unit responsible for the work was stretched far beyond its usual volume of work. More significantly, even in a period of acute state paranoia, not one of the UFOs was deemed a risk to national security.

There were other spikes in UFO sightings in 1957 when the Soviet Union launched its *Sputnik* satellite, and in 1978, when the film *Close Encounters of the Third Kind* was released. When *The X-Files* finally disappeared off our television screens in 2002, the whole idea seemed suddenly unfashionable. "The Martians aren't coming: British UFO-spotting clubs may have to close because of a lack of sightings," read one headline.

Credence of UFOs is fueled by the abundance of apparently credible witnesses, such as United Airlines pilots, but also routinely including police and military officers. One such was Peter Horsley, a one-time equerry to Prince Philip who ended his career as deputy commander-in-chief of British Strike Command. Horsley became interested in UFOs while working at Buckingham Palace in the early 1950s precisely because so many reports came from airmen. On the prince's nod, he was permitted to look into the "more credible" reports, provided there would be no publicity.[2]

He did keep his counsel at the time, but his 1997 memoir is an unintentional casebook of military psychosis. At one point, he meets a general who believes that UFOs are alien spaceships come to warn us of nuclear war. "This was heady stuff but I knew that there are always a number of senior retired officers who are attracted to all sorts of fringe cults, most likely out of boredom."[3]

It seems it takes one to know one. Horsley is soon taken to meet a mysterious figure who wants an introduction to Prince Philip, for whom he has a message about humankind's depredations against wildlife and who he believes can help him in his mission to promote "galactic harmony." Horsley recounts eleven pages of "verbatim" dialogue with this figure, full of curiously accurate details about developments in science and technology that took place from the 1950s to the 1990s.

Aside from his thoughtful personal appeal to Prince Philip, the (future) president of the (not yet founded) World Wildlife Fund, the figure's comments on our destructiveness in war were entirely characteristic of aliens' sermonizing, noted in other witness reports. It does seem a shame that aliens would come all that way to warn us of a danger of which we were already acutely aware at the time, rather than tell us something usefully prescient about HIV, CFCs, or CO_2 emissions.

From time to time, these distinguished witnesses gain sufficient momentum to challenge the establishment (people very like themselves, of course) about the cover-up they suspect to be taking place. The *Daily Express* reported one gathering in Washington, D.C. with the headline "Don't tell the CIA but generals have proof ET exists and wants to make contact." Appearing at the conference were men from the FAA, a cardinal from the Vatican, and Britain's former Chief of Defence Staff, Admiral Lord Hill Norton, who believed that "there is a serious possibility that we are being visited—and have been visited for many years—by people from outer space, from other civilisations; that it behoves us to find out who they are, where they come from, and what they want."

During Congressional hearings on UFOs in 1966, the future American president Gerald Ford asked, "Are we to assume that everyone who says he has seen a UFO's an unreliable witness?" Well, it might be a good start. For, again, which is more likely: that a military officer suffers from some delusional condition, or that he has been visited by aliens? "Credible" does not mean should-be-believed-under-all-circumstances, and even men in uniform may have funny turns. Even if we accept that a person with military training is in general a more reliable witness than an ordinary member of the public, there are large numbers of military personnel and so there is still a good chance that a few of them are experiencing psychological difficulties. Add in the special conditions that might obtain, such as forms of stress or oxygen starvation in flight, and we might be tempted to say that military UFO sightings should be given no special credence at all.

Although aliens might seem inseparable from their vehicles, the creatures themselves present their own distinctive dangers. Ufology is essentially a harmless pastime—perhaps even useful, if it diverts bored air marshals from ordinary war-making. But aliens do claim human lives and minds. In the most spectacular case of "alien abduction," 39 members of the Heaven's Gate cult were found dead in April 1997, their souls having supposedly left their bodies in order to rendezvous with a spacecraft come to take them away.

More frequently, individuals report seeing and hearing things, or discover that they cannot account for periods of time, and attribute this to alien contact or abduction. One in ten Americans claims to have seen one or more UFOs, a far higher incidence than the few hundred sightings a year investigated by the Air Force. Polls routinely find that a majority of people believe there is intelligent life on other planets, and around half "believe in" UFOs (although of course they'd be fools not to, at least while the U part still applies). But 65 percent of Americans also believe that a UFO crashed at Roswell, New Mexico, according to a CNN poll in 1997, on the fiftieth anniversary of the most celebrated claim of an alien encounter. Nearly 4 million Americans claim to have been *abducted* by (not just to have seen) aliens according to another poll in 1992. Merely as tourism this is quite a figure—it's more Americans than visit France each year.

Aliens seem to be created in our own image. They are bipedal, they possess similar although superior technologies and have similarly dismaying ambitions of domination and control. Curiously, Americans' reports over the years have gradually come to concur on the key details of what aliens look like. Equally curious, reports of aliens made by people in other cultures diverge from this norm.

This is to be expected when one considers the sources of inspiration available to these groups, while the disappointing general likeness of aliens to humans is not so much a failure of imagination, but a clue that aliens are not aliens at all but proxies for humans. Accounts of alien abduction closely resemble accounts of abuse, as Carl Sagan points out in *Demon-Haunted World*. Elaine Showalter adds that women's "abduction scenarios closely resemble women's pornography."[4] They may also reflect anxieties over conception and childbirth. The obsession with sex squares neatly with the idea of higher intelligence: of course these aliens want us for our bodies—they'd have no use for our inferior minds.

But wait. Why not genuine aliens? Consider alien visitations not as a problem in psychology, but as one in probability. How many aliens are out there, and would they find us? In 1961, Frank Drake, an astronomer at Cornell University, devised a formula to find out. The Drake equation is simply the multiplication of a set of chances. Start with the number of stars in our galaxy, the Milky Way. Take the fraction of those that have planets; take the fraction of those planets that are chemically suitable for life; take the fraction of *those* planets on which life actually does arise; now consider the fraction where that life evolves to intelligence; and then the fraction that develops appropriate communication (or interstellar travel) technology. Most of these fractions are small, some very small, so that multiplied together they amount to a very small chance indeed. Still, the number of stars in the Milky Way is several hundred billion, so that helps raise the odds. Present estimates of the number of civilizations out there potentially able to communicate with us range from effectively none to maybe 5,000.

This looks good, but now contact must be made. This brings in another series of long odds. The alien civilization must point their detection equipment our way—one planet orbiting one of several hundred billion stars. They must recognize the signal of our presence and they must then choose to act. Plus, there's one final, vital term from the Drake equation to consider—one that makes the equation a child of its time during the Bay of Pigs fiasco. This is the fraction of the lifetime of the planet during which its communicating intelligent life flourishes. Both the alien civilization and our civilization must overlap in time if we are to make contact. We have only had the ability to send communications into space for around 50 years, one ten billionth of the lifetime of the Earth. Based on the Earth experience, civilizations tend to last no more than 500 years, and perhaps it's the same on other planets. The chances of our connecting become incredibly tiny. Small wonder, then, that as John Allen Paulos cruelly puts it, "innumerates are considerably more likely than others to believe in visitors from outer space."[5]

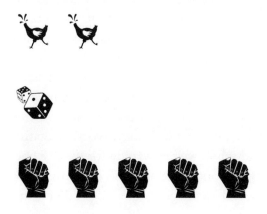

That's When It Hits You

"Living on borrowed time"
—*The New York Times*

You might want to put this date in your diary: April 13, 2029. It's a Friday. Friday the 13th. This is the day, NASA announced in 2004, on which the Earth is most likely to be struck by a civilization-destroying asteroid. On Christmas Eve 2004, the space agency quoted odds of one in 300—an unprecedented level of risk—that we would be hit by the recently discovered 2004 MN4, a 400-meter diameter chunk of rock orbiting around the sun. Later that day it dramatically shortened the odds to 1 in 63. By the end of Christmas Day, the chance of the planet being largely wiped out stood at 1 in 45. On the Torino scale, asteroid watchers' newly invented equivalent of the Richter scale, 2004 MN4 rose from a zero to a two and then to a four. These may be long odds for betting on a horse, but they are uncomfortably short when you consider what's at stake.

But at least there were 25 years to work out what to do about it.

Most asteroids are thought to be remnants of a failed planet. They generally occupy an orbit between Mars and Jupiter, but numbers of them are regularly dislodged from this orbit by the gravitational influence of the planets. Small asteroids reach us all the time but burn up in the Earth's atmosphere, where we see them as meteors or "shooting stars." A meteor that survives this process is termed a meteorite when it reaches the Earth's surface. There are presently around 3,000 designated "near-Earth" asteroids the size of 2004 MN4 or larger. "Earth will be hit by an asteroid large enough to wipe out most of the human race. That is a certain fact. We just don't know when," according to Lembit Öpik, a worried British member of parliament.[1] And while we are waiting for the big one, there are also a billion objects out there the size of a bus, quite large enough to do considerable damage to our planet. These are frightening numbers.

But consider what it must take for one of these asteroids to pose a real threat to us. First, the asteroid must be of sufficient size. Fortunately, the abundance of asteroids decreases sharply at larger sizes, so while there are indeed many out there, the billion or so bus-sized objects in fact represent a tiny minority of the total number of asteroids. Next, the asteroid must have enough energy to penetrate the atmosphere and do damage. This means that it must be both massive and fast-moving. Then, its composition must be right—a dense stony or metallic object will do more damage than a carbon-rich or icy one, which is more likely to break up high in the atmosphere.

Most obviously, the orbit of the asteroid must coincide with that of the earth. This is no small requirement. Space, as Douglas Adams pointed out in *The Hitchhiker's Guide to the Galaxy*, is "big, really big." So the chance of the path of one fairly small orbiting rock (the Earth) overlapping with that of a far smaller one (the asteroid) is always going to be extremely low. This overlap must occur not only in space but also in time. And again, the chance that our asteroid crosses the Earth's path at the precise moment that the Earth itself is at that point in its orbit is very small. In 1989, an asteroid missed the Earth by just six hours, which sounds close until you express it in terms of distance—400,000 miles, well over the distance from the Earth to the moon.

Finally, even if an asteroid does strike, it may not have a catastrophic impact on us. Only a very rare, large asteroid would have major consequences for humankind regardless of where it hit the planet's surface. Clearly, a rather smaller object could destroy a city like Los Angeles or Tokyo in a direct hit, but it is far more likely to fall in an almost entirely unpopulated region. If a meteorite falls into the sea, however, then the mortality due to the resulting tsunami is likely to far exceed that for an equivalent impact on land. A modest 300-foot asteroid might kill 10,000 people in a land impact, but the tsunami it would cause if it landed in the sea might kill 100 million. "The United States is, by this measure, one of the most vulnerable nations on Earth, since it has numerous major cities close to sea level on two separate oceans," observes the asteroid expert John Lewis.[2]

The good news is that these probabilities, each tiny on its own, must be multiplied together to calculate the overall risk. Since all the fractional chances stacked up here are extremely small, the multiplied total giving the overall probability of a lethal impact is minuscule. The practical risk could be reduced still further if we have forewarning of an impact.

Asteroid 2004 MN4—whose notoriety duly earned it a proper name, Apophis—is bigger than the meteorite responsible for the geological feature known as Meteor Crater in Arizona and far bigger than the one that exploded in 1908 with the force of a thousand Hiroshima bombs in the air above Tunguska, razing thousands of square miles of Siberian forest.

So why were there not desperate headlines at the time warning of our impending destruction? Because the danger passed almost immediately. With the odds at 1 in 300, the NASA scientists seemed strangely keen not to alert the world to the danger but to offer reassurance. "These odds are likely to change on a day-to-day basis as new data are received," they announced. "In all likelihood, the possibility of impact will eventually be eliminated as the asteroid continues to be tracked by astronomers around the world."[3]

Now, "in all likelihood" is hardly a phrase of scientific precision. What did they mean? The newspapers were certainly puzzled. "Asteroid impact alert for 2029? Perceived danger may go down as studies continue," the *Seattle Times* repeated dubiously. In fact, as we know, the odds rapidly shortened. But even then, the scientists remained apparently blasé: "The odds against impact are still high, about 60-to-1, meaning that there is a better than 98 percent chance that new data in the coming days, weeks, and months will rule out any possibility of impact in 2029."[4] This was a little bit different. But why should new data lean this way? Mightn't they equally be found to suggest that a collision is more likely?

After all, this does not happen for other statistics, although we might wish that it would. If, in a certain city, you have a one-in-a-hundred chance of being shot, that risk does not disappear as a gunman approaches you. A better analogy is one that omits human malevolence, such as emerging from a junction on to a main road in your car and having to cross traffic going the other way. You can risk the turn when you see another car approaching if that car is going to turn off the main road on to your road. As the car gets closer, you gather more data about it—you see it slow down, indicate, the wheels turn slightly, until at a certain point you are sufficiently sure it won't hit you to make your maneuver. A risk that once seemed great is in this way shown to be less.

Scientists estimate an asteroid's future trajectory based on observations of its orbit around the sun. Because of observational inaccuracies and limitations in the computer models they use, this path is not a fine line but a three-dimensional swathe of space. Where that swathe overlaps the Earth's orbit, there is a finite possibility that the asteroid will hit us. But as more data are obtained, the swathe can only become narrowed, and as this happens the likelihood that it now includes the Earth is reduced.

In the case of Apophis, somebody found some archived photographic plates of the asteroid dating from before its "discovery." Measurements from these promptly reduced the risk to zero, although closer examination later revealed a bias in the measurements which meant that the "all clear" should not have been sounded at all. Two months later, the danger was finally eliminated with new data from the Arecibo radiotelescope.

Apophis is not beaten yet, however. Its orbit crosses ours again in 2036 and it is currently given a 5,000-to-one chance of hitting us. This led the popular television astronomer Patrick Moore to predict, in an article on space exploration milestones for the twenty-first century, that in 2028 as the asteroid approaches the Earth for the first time we might send up a nuclear device to deflect it from its course so that it does not score a hit the second time around.

Asteroids are a new fear. Until recently, we simply did not know enough about them to worry. Two recent discoveries have changed our perceptions. The first was the scientific confirmation obtained in 1990 that an asteroid strike at Chicxulub in Mexico was the most likely cause of the extinction of the dinosaurs 65 million years ago. The second was the dramatic footage of the comet Shoemaker–Levy 9 breaking up and pummelling the planet Jupiter in 1994. At the same time, improved astronomical observation is enabling us to spot many more, especially smaller asteroids—and realize that they pass very close to Earth. Compare the 3,000 "near-Earth asteroids" known now to only 18 known in 1981. Methods developed in order to evaluate the impact on human populations of nuclear weapons meanwhile allow us to estimate the damage that would be done if an asteroid did hit.

Now that we are learning more, the impact risk is apparently both rising (as we find more asteroids) and falling (as we discover, one by one, that they are not after all on a collision course with Earth). There has been a large drop in the estimated interval between globally catastrophic strikes. In the nineteenth century, this was put at 281 million years. In 1958, the pioneer asteroid researcher Ernst Öpik (Lembit Öpik's grandfather) concluded that an object 1,600 feet in diameter—enough to trigger the end of civilization—could be expected every 590,000 years. This has altered to 5–10,000 years today.

These are still low probabilities of destruction, but there remains a disproportionate public fascination in them which may be explained by a number of other factors. Here is a natural hazard with no limit to the scale of its destruction—unlike more familiar terrestrial dangers of earthquakes, volcanoes, or tsunamis. "They are the only credible natural threat to human civilization," according to John Lewis.[5]

Where public alarm is far out of proportion to actual risk, it is hard to judge the effort that should be made to address the risk. Media coverage seldom helps. In 1999, BBC News warned that the Earth is "due" to be struck by a "giant asteroid" and that we must: "Invest to avert armageddon." The piece quoted Lembit Öpik, whose humble aim was for the UK to contribute $700,000 toward a global initiative to obtain better astronomical data. "If we saw an asteroid hurtling toward us," he said, "we would get 20 seconds and that's not even long enough for the Lord's Prayer. If we make this investment then we would get anything from two years' notice of an impending impact and that's long enough to divert the object." On the first working day of the new millennium, the British government announced that it would fund a task force to look into the threat, while the then science minister Lord Sainsbury soothed: "This is not something that people should lie awake at night worrying about."[6] The major outcome was to set up the so-called Near Earth Objects Information Centre.

Meanwhile, astronomers' top priority is to log the thousand-plus objects 3,000 feet or more in size that their calculations tell them must be orbiting within the solar system, even though one-third of them are strictly "undiscovered." Smaller objects are more numerous, and so it is more likely that one will hit the Earth, but the damage they do is less, and of course it is harder to spot them. The chance of any meteorite, even a small one, hitting you person-ally can be estimated by considering the fraction of the area of the planet's surface you occupy. The Earth's surface area is 300 million square miles. Seen from above, you take up, let us say, an area about 1.6 square feet. This gives a chance of 1 in 2 million billion that any meteorite hits you (or the roof above your head), and a chance of 1 in 300,000 or so that it hits someone somewhere. Perhaps 50,000 meteorites a year reach the Earth weighing 0.4 ounces or more. We would therefore expect 1 human life to be lost to meteorite strikes every 6 years on average. Why don't we hear more about these astonishing injuries and deaths? According to Lewis, the reason is that the death is often erroneously put down to more plausible natural mishaps or explained by superstitious beliefs.

Suppose the worst: A big asteroid is found that really does have our name on it. Even then, the risk "can be mitigated in much more concrete ways than is true of most hazards," according to Clark Chapman of the Southwest Research Institute in Boulder, Colorado. "An impact can be predicted in advance in ways that remain imperfect but are much more reliable than predictions of earthquakes or even storms."[7] With ample warning, people could be moved away from the area of impact.

Launching a nuclear bomb into space to deflect an asteroid from its course is an option under serious consideration. But this carries its own risks. Some regard the supporters of this technological fix as fanatics. In 1994, Carl Sagan warned against the diversion "through error or madness" of a premature nuclear bomb space mission "for other, nefarious, purposes."[8] The question we should ask ourselves is this: Is the risk of our destruction through such misuse greater than that due to the undeflected asteroid?

An even greater danger, according to Chapman, comes not from a real asteroid but from panic spread by misleading reports of obser-vations or predictions. News of a near-miss by a small asteroid given at short notice would be sure to stoke public apprehension of a real impact, for example. Reports of a scientific prediction (which later proves wrong or to have been misreported) that an impact will occur at a particular time and place would cause intense panic until (and perhaps after) the report was denied.

In the past, such false alarms have been avoided because scientists have kept apparently bad news to themselves until they had information to say that it was not news at all. In future, it is increasingly likely that these alerts will leak out, and we will have to learn to make our own judgments. It is perhaps worth noting, therefore, that at the time of writing, all asteroids known to NASA rank as zero on the Torino scale.[9]

A Skeptic's Toolkit

For those not inclined to credulousness, it is easy to be cynical about the media in its presentation of stories involving statistical or scientific information. We would rather encourage skepticism. With this in mind, we offer this toolkit for the interpretation of data or information that come our way.

- **Vested interest:** Ask yourself who has made a particular statement. Why might they have done this? Are we being told the whole story?
- **Weasel words:** These should ring alarm bells—especially emotive ones such as "plague," or ones that put us on a one-way trip to disaster such as "inevitable" and "overdue." It is inevitable that night follows day, but it is not inevitable that there will be a terrorist attack. You can be overdue for a meeting that started an hour ago, but a volcanic eruption, an earthquake, or an outbreak of disease is only ever overdue based on arguments of probability. Other words may not have the obvious meaning. Government surveys of the "work force" count anyone who has worked one hour or more in a week, so a boost in the numbers working could be down to children babysitting or students spending an evening behind a bar. Is this what you consider work?
- **Surveys:** Who conducted it? Are they credible? Do they have an obvious motive? Who paid them? Whole fields of study can become unhealthily dependent on funds from one source, whether that source is commercial, governmental, or charitable. Were the questions neutrally worded? How big is the sample? Too small, and the result may be skewed; too big, and the authors may be trying to use sheer weight of numbers to persuade you. Is the sample size and margin of error shown? When a pet food

manufacturer says that four out of five cats prefer their product, did they only feed five cats? How were the data collected?

- **Figures:** Try to compare figures. Look at as many of the effects of a change as possible, not just one. Compare the present with the past. Compare one country with another. If the data aren't there to make the obvious comparison, ask yourself what is being obscured.
- **Percentages and actual numbers:** People with a story to tell will choose the more impressive way of putting things.
- **Anecdote and statistics:** Fears are spread by word of mouth, press, and television reports based on harrowing individual stories; authorities frequently counter these with broad statistics. Meaningful comparison between the two is hard. Both may be "true."
- **Graphs and charts:** Like words and figures, these may be subject to error or deliberate distortion. Don't automatically believe them because they look technical.
- **Timeframe:** This is an important factor that words like "inevitable" gloss over. Sea levels are rising, but over a longer period than housing planning cycles, so there is time to adapt. Many data series have a long-run trend, a shorter cyclical variation, and then (often erratic) individual data points. Be aware of each so as not to be tricked.
- **Why now:** Ask yourself why the story is appearing now, and whether it would be equally newsworthy at another time. Global warming stories appear more in the summer. Travel fears play well as people set off on their holidays. Sex surveys are often released in time for Valentine's Day.
- **Defeatism:** Be wary when told there is nothing we can do about something. Why then are we being told about it? Is it merely to alarm us, or to put us in a state of fear?
- **Scare snobs:** Distrust scares where an elite is trying to deny others advantages they already enjoy, for example environmental and health crises exacerbated by cheap flights, exotic food, private modes of transport, choice in medicine and education.

- **Scenarios:** Many economic and scientific studies model a range of future scenarios. Make sure that the outcome described is not just the worst-case scenario.
- **Accentuate the positive:** Don't discount the possibility that even if some things are getting worse, others may get better—which negative newspaper stories make it their business to do. It will get warmer in 100 years, but what might human ingenuity have devised by then? New energy sources? Genetically modified human metabolism? Improved photosynthesis? Science fiction, you might say, and so it is—for now. But think what has been achieved over the last 100 years.
- **The big picture:** It's bad if 100 people die of bird flu, but in a country of 50 million, this is very few. How many died of everything else?
- **A sense of proportion:** Try to keep one, even if the top brass won't. Germany's Interior Minister Wolfgang Schaeuble, insists that Islamic terrorism is the single largest threat to Germany's stability. Does this seem remotely credible, or is somebody just bigging himself up?

Franklin Delano Roosevelt was not wrong when, at the deepest point of the Depression in March 1933, he gave his first inaugural address, saying, "The only thing we have to fear is fear itself."

Notes

Introduction

1. Andrew Marr, *My Trade: A Short History of British Journalism* (London: Macmillan, 2004).
2. Charles Mackay, *Extraordinary Popular Delusions and the Madness of Crowds* (New York: Three Rivers Press, 1980), p. 266.
3. www.psandman.com.
4. John Allen Paulos, *Innumeracy: Mathematical Illiteracy and Its Consequences* (New York: Hill and Wang, 2001).
5. According to the quotations expert Nigel Rees, wwwic. btwebworld.com/quote-unquote/p0000149.htm.
6. *The Times*, quoted in *RSS News*, June 2006.
7. http://en.wikipedia.org/wiki/GeorgeCarlin.
8. "Living dangerously: a survey of risk," *The Economist*, January 24, 2004, p. 8.
9. "Age of terror scaring Australian children," *The Age*, September 26, 2006.
10. Mary Douglas and Aaron Wildavsky, *Risk and Culture: An Essay on the Selection of Technical and Environmental Dangers* (Berkeley and London: University of California Press, 1982), p. 9.
11. Mackay, op. cit., p. xv.

Chapter 1. Sex, Marriage and Children

THE BIRTH DEARTH

1. "Italian women shun mamma role," www.bbc.co.uk, March 27, 2006.
2. Donella H. Meadows, Dennis Meadows, Jorgen Randers, and William Behrens, *The Limits to Growth: A Report for the Club of Rome's Project on the Predicament of Mankind* (New York: Potomac Associates, 1972).
3. Paul Ralph Ehrlich, *The Population Bomb* (New York: Ballantine, 1970).
4. "World population prospects," www.un.org.

FAMILY BREAKDOWN

1. Hillary Rodham Clinton, *It Takes a Village* (New York: Simon & Schuster, 2006).
2. "It Takes a Family," *The Washington Post*, July 25, 2005.
3. "Parents live apart to cash in on benefits system," *Daily Telegraph*, December 16, 2005.
4. "Survey: Working Dads Want More Family Time," www.cnn.com, June 14, 2007.
5. Adecco Parents in the Workplace Survey 2008.
6. "After-school worries: tough on parents, bad for business," Brandeis University report, December 2006.
7. UNICEF, *Child Poverty in Perspective: An Overview of Child Well-being in Rich Countries*, UNICEF Report Card 7, February 14, 2007 (UNICEF Innocenti Research Centre, 2007), available at www.unicef-icdc.org/presscentre/presskit/reportcard7/rc7eng.pdf.
8. "Marriage rates fall to lowest on record," www.statistics.gov.uk, February 21, 2007.
9. "Marriage and divorce," www.cdc.gov.
10. "Trends in characteristics of births by state: United States, 1990, 1995, and 2000–2002," www.cdc.gov, May 2004.

11.   "Mamma's boys," *Observer*, November 12, 2006.

12.   "Do divorcing couples become happier by breaking up?" *Journal A*, Royal Statistical Society, March 2006.

13.   "Pregnancy rate drops for U.S. women under age 25," www.cdc.gov, April 2008.

14.   "Marriage maketh man," *The Times*, March 1, 2007.

THE MARRIAGE SQUEEZE

1.   "A good man is harder to find," *Newsweek*, June 2, 1986.

2.   "Marriage by the numbers," *Newsweek*, June 5, 2006.

3.   "An iconic report 20 years later," *Wall Street Journal*, May 25, 2006.

4.   *Home Alone?* Unilever Family Report 2005, www.ippr.org.uk, October 27, 2005.

5.   "It's worth the wait," *Sunday Times*, June 4, 2006.

6.   "Forget Bridget Jones—meet Brad," www.bbc.co.uk, November 9, 2004.

SOMETHING FOR THE WEEKEND?

1.   BRG Consult at www.consultgb.com.

2.   www.durex.com/cm/gss2005results.asp.

3.   "Much more sex please . . . we're British," *Guardian*, August 18, 2004. Other variants on the title of the theatrical farce *No Sex Please, We're British* include the mutually contradictory "No quick sex please, we're British" and "No sex please, we're busy."

4.   http://www.tennessee.gov/humanserv/cs/wtr/wtr.htm.

5.   Anne M. Johnson et al., "Sexual behaviour in Britain: partnerships, practices, and HIV risk behaviours," *Lancet* 358 (December 1, 2001), pp. 1835–42; Kaye Wellings et al., "Sexual behaviour in Britain: early heterosexual experience," *Lancet* 358 (December 1, 2001), pp. 1843–54.

6.   Kaye Wellings et al., "Sexual behaviour in context: a global perspective," *Lancet* 368 (November 11, 2006), pp. 1706–28.

Chapter 2. Health

THE FAT THING

1.   "Overweight and obesity: defining overweight and obesity," www.cdc.gov, October 2006.
2.   "Overweight and obesity: home," www.cdc.gov, October 2006.
3.   "Quality and outcomes framework guidance: obesity," www.bma.org.uk, February 2006.
4.   *Independent*, April 22, 2006.

CURRYING FLAVOR

1.   "Is your curry killing you?" *Sun*, February 2, 2006.
2.   Hervé This, *Molecular Gastronomy: Exploring the Science of Flavor* (New York and Chichester: Columbia University Press, 2006), p. 94.
3.   "A pinch of controversary shakes up dietary salt," www.fda .gov, quoting Alicia Moag-Stahlberg.
4.   Harold McGee, *On Food and Cooking: The Science and Lore of the Kitchen* (New York: Scribner, 2004).
5.   www.smartmobs.com/archive/2006/07/07/qr_code_ hyper.html.
6.   Joanna Blythman, *Bad Food Britain* (London: Fourth Estate, 2006), p. 172.
7.   "Salt debate leaves bitter taste," www.foodproductiondaily .com/redirect.asp?idc=13228, March 23, 2006.
8.   Joanna Blythman, *The Food We Eat* (London: Michael Joseph, 1996), p. 258.
9.   www.italkali.com/english/home.htm.

10. "Learn to shake the salt habit," www.msnbc.com, May 23, 2008.

A DEAD DUCK

1. Mike Davis, *The Monster at Our Door* (New York: New Press, 2005), p. 5.
2. www.who.int/csr/disease/avian_influenza/country/cases_table_2007_10_08/en/index.html.
3. Davis, op. cit., pp. 76, 8.
4. Quoted in ibid., p. 125.
5. David Boyle, *The Tyranny of Numbers: Why Counting Can't Make us Happy* (London: HarperCollins, 2000), p. 80.

IT'S AMAZING WHAT THEY CAN DO

1. Laurie Garrett, *Betrayal of Trust: The Collapse of Global Public Health* (Oxford: Oxford University Press, 2001), p. 256.
2. Geoffrey Cannon, *Superbug* (London: Virgin, 1995), p. xxii.
3. Garrett, op. cit., p. 248.
4. Garrett, op. cit., p. 258.
5. Paul Slovic, *The Perception of Risk* (London: Earthscan, 2000), p. 92.

COMPLETING THE COURSE

1. www.vaccineinformation.org
2. "Measles—United States, first 26 weeks, 1989," www.cdc.gov, December 22, 1989.
3. A. B. Bloch et al., "Health impact of measles vaccination in the United States." *Pediatrics 76 (4):* pp. 524–32.
4. A. J. Wakefield et al., "Ileal-lymphoid-nodular hyperphasia, non-specific colitis, and pervasive developmental disorder in children," *Lancet* 351 (February 28, 1998), pp. 637–41.
5. www.cochrane.org/reviews/en/ab004407.html.

6. Committee on Safety of Medicines, Report of the Working Party on MMR Vaccine, 1999.
7. Stanley Feldman and Vincent Marks, *Panic Nation* (London: John Blake Publishing, 2005), p. 109.
8. "Health, United States, 2007," www.cdc.gov, p. 257.

SUDDEN DEATH

1. http://www.cdc.gov/sids/suid.htm.

Chapter 3. Passing the Time

ART IS DANGEROUS

1. *L.A. Life*, October 8, 1992.

CHEERS!

1. "The unbearable lightness of being English," *New Zealand Listener*, January 15, 2005.
2. www.icap.org.
3. "London women are shameless drunks," *Daily Mail*, April 30, 2003.
4. "Quarter of children so drunk they have passed out," *Daily Express*, November 17, 2005.
5. http://www.cdc.gov/HealthyYouth/yrbs/index.htm.
6. "Alcohol in Europe: a public health perspective," www.ias .org.uk, June 1, 2007.
7. http://www.drugabuse.gov/EconomicCosts/Chapter1 .html#1.8.
8. "The not-quite-so-grim neurology of teenage drinking," www.stats.org, July 7, 2006.
9. "Who is minding the Internet liquor store?" NBC News, August 9, 2006.

10.   "Selling alcohol online survey snares NBC," www.stats.org, August 16, 2006.

11.   "Drink data to show disorder fall," *Financial Times*, February 6, 2006.

THE DEATH OF CINEMA

1.   Paulo Cherchi Usai, *The Death of Cinema: History, Cultural Memory and the Digital Dark Age* (London: British Film Institute, 2001).

Chapter 4. Social Policy

GOLDEN OLDIES' TIME BOMB

1.   "Pensions crisis is much worse than firms say," *Scotland on Sunday*, November 19, 2006.

2.   Europe Pensions Barometer 2006, AON Consulting, www.aon.co.uk, January 2007.

3.   'The new pension crisis," *Wall Street Journal*, August 18, 2006.

CREDIT CRUNCH

1.   "The iPod generation," *Daily Telegraph*, September 18, 2006.

2.   "Graduates fear debt more than terrorism," *USA Today*, May 18, 2005.

3.   OECD Economic Outlook no. 80, www.oecd.org, November 28, 2006.

4.   "Money sickness syndrome," www.axa.co.uk, January 20, 2006.

5.   OECD Economic Outlook, op. cit.

THE HOUSING BUBBLE

1. "Global house prices are still rising," *Financial Times*, December 30, 2006.
2. "House Price Increases Continue; Some Deceleration Evident," www.ofheo.gov, June 1, 2006.
3. International housing affordability survey, Demographia, 2006.
4. "The UK house-price-bubble illusion," www.housingoutlook.com
5. www.johnkay.com.

IMMIGRANT INVASION

1. From a speech to Portland State University.
2. Patrick J. Buchanan, *State of Emergency* (New York: Thomas Dunne Books, 2006), pg. 221.
3. "Environmental migration," Optimum Population Trust, November 2, 2006.
4. "International migration outlook," 2007, OECD.
5. Ibid.
6. "Foreign labour in the UK," *Labour Market Trends* (ONS, October 2006).

LOSING CONTROL OF YOUR VEHICLE

1. "Economic impact of U.S. motor vehicle crashes reaches $230.6 billion, new NHTSA study shows," www.dot.gov, May 9, 2002.
2. Paul Slovic, *The Perception of Risk* (London: Earthscan, 2000), p. 140.
3. www.bts.gov.
4. www.dft.gov.uk/pgr/statistics/datatablespublications/accidents/casualtiesgbar/roadcasualtiesgreatbritain2005.
5. www.rotor.com.

6. "A Rash of Medical Helicopter Crashes Brings Call for Reform," *Wall Street Journal*, October 29, 2008.
7. Slovic, op. cit., p. 26.
8. Cited in Simon Briscoe, *Britain in Numbers* (London: Politico's, 2005), p. 193.
9. ibid.

DEATH BY PHONE

1. *The Register*, December 1, 2003, www.theregister.co.uk.
2. "Association between cellular telephone calls and motor-vehicle collisions," *New England Journal of Medicine*, February 13, 1997.
3. www.bbc.co.uk, March 22, 2002.
4. *Daily Mail*, November 4, 2005.
5. "Mobile phones and driving," www.dft.gov.uk.
6. "Cost effectiveness of regulations against using a cellular telephone while driving," *Medical Decision Making* 19(1) (1999).

Chapter 5. The Workplace

NO WORK OR LOW PAY

1. "Don't discount the positive side of globalization," *Toronto Star*, December 31, 2006.
2. "And now for a word about globalization," *Arizona Republic*, November 9, 2006.
3. "Profits of doom," *Financial Times*, October 14, 2006.
4. "Economic challenges facing the middle class," Congressional testimony, January 31, 2007.

UNDERPAID WOMEN

1. "Memo to John Roberts," Institute for Women's Policy Research, September 2005.
2. "The gender pay gap," *Centrepiece*, 2006.
3. http://www.now.org/issues/economic/factsheet.html.
4. "Just pay", 2001, and "Britain's competitive edge," 2004, www.eoc.org.uk.
5. "Scant progress on closing gap in women's pay," *New York Times*, December 24, 2006.
6. "Women suffer pay gap," *Daily Mail*, January 17, 2006.
7. "A third of women banking on men," *Metro*, July 11, 2006.
8. "Study: Women concentrated in 'pink collar' jobs," www.blr.com (Business & Legal Reports), May 6, 2003.
9. "Rise of the women who earn more than their men," *Daily Mail*, November 15, 2006.
10. "Progress on gender pay gap stalled," Fawcett Society, October 21, 2006.
11. "30 years on," Fawcett Society, November 7, 2005.
12. "Shaping a fairer future," Prosser Report, Women and Work Commission, February 2006.

IT'S ALL TOO MUCH

1. "Stress a major health problem in the U.S., warns APA," www.apa.org, October 24, 2007.
2. http://www.stress.org/job.htm (The American Institute of Stress).
3. "Call centres under pressure," www.channel4.com, April 2005.
4. http://ki.se, October 9, 2006.
5. "Work related ill-health affects 2 million," *Financial Times*, August 4, 2006.
6. "Stress and coronary heart disease," National Heart Foundation of Australia, www.mja.com, March 17, 2003.

7. "A health warning for older armchair fans," *Financial Times*, June 30, 2006.

8. "In-flight confrontations can lead to charges defined as terrorism," *The Los Angeles Times*, January 20, 2009.

9. "In the driver's seat," *Auto Vantage*, May 16, 2006.

10. "Is teaching the most stressful job?" www.bbc.co.uk, March 27, 2004.

11. www.isma.org.uk, www.ismabrasil.com.br, and www.isma-usa.org.

12. www.hypnosisdownloads.com.

13. "Ten steps to happiness," *Daily Mail*, June 28, 2006.

GAMES OF CHANCE

1. Edward Tenner, *Why Things Bite Back: New Technology and the Revenge Effect* (New York: Vintage, 1997), p. 222.

2. Stanley Feldman and Vincent Marks, *Panic Nation* (London: John Blake Publishing, 2005), p. 239.

3. *Guardian*, December 8, 2005, www.chiropractic-uk.co.uk.

Chapter 6. Law and Order

TERROR ALERT

1. "Transatlantic trends," *Key Findings 2006*, September 6, 2006.

2. www.synovate.com/knowledge/infact/issues/200510, October 2005.

3. "Similar levels of fear of terrorism in USA and Great Britain," www.harrisinteractive.com, February 6, 2004.

4. *New York Times*/CBS News Poll, www.nytimes.com, September 7, 2006.

5. NABE Economic Policy Survey, www.nabe.com, August 28, 2006.

6. "Worldwide poll shows 60 percent fear terror threat is worse after war," *Guardian*, February 28, 2006.
7. YouGov poll, quoted in *Daily Telegraph*, August 25, 2006.
8. NCTC Report on incidents, 2006, www.nctc.gov.
9. "Deaths by age, sex and underlying cause 2004," HSQ 26, www.statistics.gov.uk, February 2007.
10. "MI5 head warns of 1600 terror plotters," *Financial Times*, November 10, 2006.
11. "Secret report: terror threat worst since 9/11," *Sunday Telegraph*, February 25, 2007.

BANG BANG

1. AKUF and the University of Hamburg, quoted in *Vital Signs 2006–2007*, Worldwatch Institute.
2. Heidelberg Institute for International Conflict Research, quoted in *Vital Signs*, op. cit.
3. "Mortality after the 2003 invasion of Iraq," www.lancet.com, October 11, 2006.
4. www.iraqbodycount.org.
5. "Afghanistan risk exposed," *New Scientist*, September 9, 2006.
6. www.oecd.org/dac.
7. "One In Five Iraq and Afghanistan Veterans Suffer from PTSD or Major Depression," www.rand.org, April 17, 2008.
8. "Reflections on Gulf War Illness," *Phil. Trans. of the Royal Society*, March 24, 2006.
9. "Veterans told: there is no Gulf syndrome," *The Times*, March 25, 2006.

CRIME & PUNISHMENT

1. www.wrongdiagnosis.com.

2. "Easier to panic than do the math," *Toronto National Post*, April 13, 2006.
3. *Sunday Telegraph*, September 17, 2006.
4. "Going off track," *Philadelphia Daily News*, July 10, 2006.

Chapter 7. The Natural World

LOOKING UP

1. Al Gore, *Earth in the Balance* (Boston: Houghton Mifflin, 1992), p. 88.
2. "Late lessons from early warnings: the precautionary principle 1896–2000" (European Environment Agency, 2001), p. 80.
3. John Gribbin, *The Hole in the Sky* (New York: Bantam, 1988).
4. G. Braathen, *Geophysical Research Abstracts* 8, 09861, 2006.
5. *Chem@Cam*, summer 2006, p. 16.
6. www.esrcsocietytoday.ac.uk/ESRCInfoCentre/PO/releases/2002/September/public.aspx.
7. Fred Pearce, *The Last Generation: How Nature Will Take Her Revenge for Climate Change* (Toronto: Key Porter, 2007).
8. J. Firor, *The Changing Atmosphere: A Global Challenge* (New Haven and London: Yale University Press, 1990), p. 43.

THE SHORT, HOT SUMMER OF 2006

1. John Houghton, *Global Warming: The Complete Briefing*, 3rd edn (Cambridge: Cambridge University Press, 2004), p. 8.
2. Jonathon Porritt, *Playing Safe: Science and the Environment* (London: Thames & Hudson, 2000), p. 63.
3. Fred Pearce, *The Last Generation: How Nature Will Take Her Revenge for Climate Change* (Toronto: Key Porter, 2007).
4. "Katrina and the waves," *Independent*, August 23, 2006.

5. Roger A. Pielke, "Disasters, death and destruction: making sense of recent calamities," *Oceanography* 19(2) (June 2006), pp. 138–47.

BECOMING UNSETTLED

1. Searched using the Factiva database, the phrase "global warming" was used 16,755 times in June–August 2006, 10,547 times during the same months of 2005, and 6,196 times in 2004. Curiously, global warming is mentioned consistently less in the winter months. The comparable figures for December–February are 10,632, 8,411 and 5,055 respectively.

2. *Warm Words*, Institute of Public Policy Research, August 2006.

3. James Lovelock, *The Revenge of Gaia* (New York: Basic Books, 2006), pp. 3–4, 7.

4. *Daily Telegraph*, August 28, 2006; *Eastern Daily Press*, July 28, 2006.

5. Climate change protagonists sometimes argue that this renegade rump has dwindled to nothing, but see, for example, a letter from 41 scientists in the *Sunday Telegraph*, April 23, 2006.

6. Quoted in Aaron Wildavsky, *But Is It True? A Citizen's Guide to Environmental Health and Safety Issues* (Cambridge, Mass.: Harvard University Press, 1995).

7. Quoted in Fred Pearce, *The Last Generation: How Nature Will Take Her Revenge for Climate Change* (Toronto: Key Porter, 2007).

8. John Houghton, *Global Warming: The Complete Briefing*, 3rd edn (Cambridge: Cambridge University Press, 2004), pp. 143–4.

9. "The threat to the planet," *New York Review of Books*, July 13, 2006.

10. S. Pacala and R. Socolow, "Stabilization wedges: solving the climate problem for the next 50 years with current technologies," *Science* 305 (2004), pp. 968–72.

PIGS MIGHT SWIM

1. Mark Lynas, *High Tide: News from a Warming World* (London: Flamingo, 2004), p. 83.
2. Intergovernmental Panel on Climate Change, *Climate Change 2001: The Scientific Basis* (Cambridge: Cambridge University Press, 2001), p. 642.
3. Lynas, op. cit., p. 232.
4. Ibid., p. 114.
5. J. D. Orford and R. W. G. Carter, "Examination of mesoscale forcing of a swashaligned gravel-barrier," *Marine Geology* 126 (1995), pp. 201–11.
6. Lynas, op. cit., p. 114.
7. See, for example, E. Rignot and P. Kanagaratnam, "Changes in the velocity structure of the Greenland ice sheet," *Science* 311 (February 17, 2006), pp. 986–90.
8. "For my people, climate change is a matter of life and death," *Independent*, September 15, 2006.

GO WITH THE FLOW

1. R. I. Tilling and P. W. Lipman, "Lessons in reducing volcano risk," *Nature* 364 (July 22, 1993), pp. 270–80; www.undp.org/bcpr/disred/documents/publications/rdr/english/c2/b.pdf; www.usgs.gov/newsroom/article.asp?ID=202.
2. "After the volcano," *Guardian*, July 18, 2005.
3. "Prophet of doom," *Independent*, November 9, 2005.
4. "Tokyo prepares for 'the big one,'" *Financial Times*, January 17, 2005.
5. "AD79 and all that," *Independent* (foreign edn), October 2, 2003.

CHILLING NEWS

1. Quoted in Aaron Wildavsky, *But Is It True? A Citizen's Guide to Environmental Health and Safety Issues* (Cambridge, Mass.: Harvard University Press, 1995), p. 370.

2. Lowell Ponte, *The Cooling* (New York: Prentice-Hall, 1976).

3. John Gribbin, *Hothouse Earth: The Greenhouse Effect and Gaia* (New York: Bantam, 1990), p. 13.

4. John Gribbin, *Future Weather and the Greenhouse Effect* (New York: Delacorte, 1982), p. 107.

5. Quoted in Robert L. Park, *Voodoo Science* (New York: Oxford University Press, 2000), p. 33.

6. Bill McGuire, *A Guide to the End of the World* (New York: Oxford University Press), p. 68.

Chapter 8. Our Declining Resources

WILD TALK

1. www.iucnredlist.org/info/tables/table1.

2. *Millennium Ecosystem Assessment* (World Resources Institute, 2005).

3. www.panda.org/newsfacts/publications/livingplanetreport/livingplanetindex/index.cfm.

4. http://www.wri.org/publication/content/8202.

5. Ibid.

6. Cited in, for example, Bjørn Lomborg, *The Skeptical Environmentalist* (New York: Cambridge University Press, 2001).

THE COD DELUSION

1. Boris Worm et al., "Impacts of biodiversity loss on ocean ecosystem services," *Science* 314 (November 3, 2006), pp. 787–90.
2. Tim Lang and Michael Heasman, *Food Wars: The Global Battle for Mouths, Minds and Markets* (London: Earthscan, 2004), p. 244.
3. "Seafood could disappear by 2048," *Chicago Tribune*, November 3, 2006.
4. Mark Kurlansky, *Cod: A Biography of the Fish that Changed the World* (New York: Penguin Books, 1997), p. 158.

NOT A WORD

1. "Language cull could leave people speechless," *Guardian*, May 25, 2002.
2. "Dying dialect gets a voice," *Houston Chronicle*, November 27, 2006.
3. "Like ancient forests displaced by houses," *The Times*, February 24, 2007.
4. *Houston Chronicle*, op. cit.
5. "France looks to the law," www.telegraph.co.uk, October 25, 2000.

Chapter 9. Modern Science

FRANKENSTEIN FOODS

1. Stanley Feldman and Vincent Marks (eds.), *Panic Nation* (London: John Blake Publishing, 2005), p. 160.
2. Frank Furedi, *Culture of Fear*, rev. edn (New York: Continuum, 2002), p. 174.
3. www.cabinetoffice.gov.uk/strategy/work_areas/gm_crops/index.asp.

4. Emma Hughes and Jenny Kitzinger, *Framing Genetic Research*, BA Festival of Science, Norwich, 2006.
5. Feldman and Marks, op. cit., p. 153.

LITTLE WONDER

1. www.its.caltech.edu/-feynman/plenty.html.
2. As of August 2007, the list includes 502 products: www .nanotechproject.org/index.php?id=44.
3. "Safe handling of nanotechnology," *Nature* 444 (November 16, 2006), pp. 267–9.
4. "Welcome to the world of nano foods," *Observer*, December 16, 2006.
5. "Small is hazardous," *Independent on Sunday*, July 11, 2004.
6. "After illness, a closer look at nano science," *Philadelphia Inquirer*, April 14, 2006.

EXPOSED

1. "The radioactive spy," *Guardian*, November 25, 2006.
2. "Three people exposed to spy radiation are sent for urgent tests," *Evening Standard*, November 27, 2006.
3. "Cities are swept for 'dirty' bombs," *Daily Telegraph*, January 8, 2004.
4. Paul Slovic, *The Perception of Risk* (London: Earthscan, 2000), p. 266.
5. Ibid., p. 267.
6. www.world-nuclear.org/info/info6app.htm.
7. Scott Lash, Bronislaw Szerszynski, and Brian Wynne (eds.), *Risk, Environment and Modernity* (London: Sage, 1996), p. 64.
8. Robert L. Park, *Voodoo Science* (New York: Oxford University Press, 2000).

9. William Stewart (ed.), *Mobile Phones and Health: A Report from the Independent Expert Group on Mobile Phones* (IEGMP, 2000), p. 3.

Chapter 10 They're Coming to Get You

EXPECTING VISITORS

1. "In the sky! A bird? A plane? A . . . UFO?," *Chicago Tribune*, January 1, 2007.
2. Peter Horsley, *Sounds from Another Room* (Barnsley: Pen and Sword Books, 1997), p. 172.
3. Ibid., p. 181.
4. Elaine Showalter, *Hystories: Hysterical Epidemics and Modern Culture* (New York: Columbia University Press, 1997), p. 196.
5. John Allen Paulos, *Innumeracy: Mathematical Illiteracy and Its Consequences* (New York: Hill and Wang, 2001).

THAT'S WHEN IT HITS YOU

1. "Space evaders," *Guardian Unlimited*, September 29, 2004.
2. John S. Lewis, *Comet and Asteroid Impact Hazards on a Populated Earth* (San Diego and London: Academic Press, 1999), p. 132.
3. www.space.com/scienceastronomy/solarsystem/asteroid_fears_020326-1.html.
4. www.space.com/scienceastronomy/asteroid_risk_041224.html.
5. Lewis, op. cit., p. 146.
6. www.space.com/news/uk_asteroid_000104_org.html.
7. Clark Chapman, "The hazard of near-Earth asteroid impacts on Earth," *Earth and Planetary Science Letters* 222 (2004), pp. 1–15.

8. C. Sagan and S. J. Ostro, "Dangers of asteroid deflection," *Nature* 368 (1994), p. 501.

9. http://neo.jpl.nasa.gov/risk/.